Tensors For Inquiring Minds

Tensors for Inquiring Minds takes readers on a journey to discover tensors – powerful mathematical concepts used in many areas of modern science.

Starting from the familiar ground of numbers and operations with numbers, readers are invited to re-examine these ideas in a new light, slowly building towards a deep understanding of the advanced mathematical concepts which they underpin. The overriding goal of this book is to explain tensors, specifically, by showing them in action and in relation to less complicated concepts, such as numbers and vectors.

Features
- Requires minimal mathematical pre-requisites beyond high-school algebra
- Written in an accessible, engaging style
- Full-color illustrations
- Numerous exercises for every chapter, including full solutions

Yury Deshko holds a Master's in theoretical physics (with a focus on general relativity) from Belarus State University and a doctorate in physics from the Graduate Center of the City University of New York (CUNY). His research at CUNY focused on experimental spectroscopy and diamond photonics. He served as a research associate at City College of New York until joining the semiconductor industry as a photonics engineer. Yury Deshko's pedagogical passion led him to teach, for a number of years, the introduction to the special theory of relativity in summer schools of Johns Hopkins University and Columbia University.

Tensors For Inquiring Minds

Yury Deshko

CRC Press
Taylor & Francis Group
Boca Raton London New York

CRC Press is an imprint of the
Taylor & Francis Group, an **informa** business

A CHAPMAN & HALL BOOK

First edition published 2026
by CRC Press
2385 NW Executive Center Drive, Suite 320, Boca Raton FL 33431

and by CRC Press
4 Park Square, Milton Park, Abingdon, Oxon, OX14 4RN

CRC Press is an imprint of Taylor & Francis Group, LLC

ISBN: 978-1-041-02644-0 (hbk)
ISBN: 978-1-041-00713-5 (pbk)
ISBN: 978-1-003-62036-5 (ebk)

DOI: 10.1201/9781003620365

Typeset in Latin Modern font
by KnowledgeWorks Global Ltd.

Publisher's note: This book has been prepared from camera-ready copy provided by the authors.

To Mom
With a Love
That
Only a Son
Can Have
♡
Iryna Dziashko
1962-2025

Contents

Acknowledgments

My sincere gratitude goes to all reviewers of the early drafts of the book for their valuable feedback. In particular, I want to thank Dr. Alex Rylyakov, Dr. Mikhail Makouski, Prof. Anton Kananovich, and Dr. Mohammad Teimourpour. To Dr. Teimourpour I must give separate thanks for numerous discussions, helpful suggestions on material presentation, and hospitality.

My friends, discussing the book with you was both illuminating and fun.

Finally, special acknowledgment must be given to my son, Daniel, for his help with fixing colors in many figures.

Yury Deshko
Weehawken, New Jersey

Preface

One of the greatest powers of the human mind is its capacity for *abstraction*. This capacity allows us to create and appreciate art, understand concepts such as happiness, duty, and so on, and to create and use mathematical concepts as practical tools of immense power and beauty.

This book explores *tensors* – a type of mathematical objects that extends the notion of numbers and vectors. The method of exploration is deliberately chosen to resemble a journey. Starting from familiar grounds of numbers and operations with numbers, a reader will re-examine familiar concepts in a new light and then will arrive at new concepts gradually, connecting the dots along the way.

Although the topic of the book is mathematical, the exploration will lack proper mathematical rigor, aiming instead at simplicity, clarity, and the use of helpful analogies.

This book is **not intended** to substitute more serious textbooks on linear algebra or tensor algebra. Hopefully, the main benefit of reading this book – either before, or after, or in addition to other books on the subject – is that it should help lower the *"mental barrier"* we all encounter when learning new concepts, especially abstract mathematical concepts.

To comprehend and enjoy the material of this book, the reader should have a solid knowledge of basic high-school algebra and an open and inquiring mind. The book is a bit longer than it could have been because all derivations are detailed and all exercises are fully solved.

Some sections are marked with an asterisk, for example, **Transposition***. Those sections contain material that is either optional or a bit more advanced than usual. These sections can be skipped without significant impact on the main message of the book.

About the Author

Dr. Yury Deshko is an American physicist, educator, and writer. He is the author of *"Special Relativity For Inquiring Minds"* – a textbook designed for undergraduate students and motivated high-schoolers.

When not working on applied physics problems in silicon photonics, Yury develops and teaches modern physics courses (Special Relativity and Quantum Physics) in summer schools for young aspiring physicists.

CHAPTER 1

Introduction

M ATHEMATICAL METHODS PLAY INCREASINGLY
important roles in many domains of science. Modern mathematical tools are numerous and require serious effort to master. The algebra
and calculus of *tensors* are good examples of this.

The goal of this book is to explain tensors by showing them in action
and in relation to less complicated mathematical objects, such as vectors and numbers. Understanding numbers and vectors is essential for
understanding tensors; therefore, the former two concepts are discussed
in details first.

The development of concepts will happen in the following direction:

$$\text{Numbers} \quad \rightarrow \quad \text{Vectors} \quad \rightarrow \quad \text{Tensors}.$$

We will start with reviewing numbers as the simplest mathematical objects, and will consider operations on numbers – functions – in a more
general and abstract way than one usually does in school. Many abstract
concepts related to numbers and functions will be useful for studying
vectors and tensors.

From numbers and numeric functions, we will move on to vectors.
Vectors are closely related to numbers and can't even be properly defined without the latter. Vectors are more powerful than numbers and
represent the next step in the hierarchy of mathematical objects. Vectors and functions on vectors (operations or *operators*) provide many
new concepts that are crucial for understanding tensors.

Careful study of vectors and functions on vector (operators) will in-
evitably lead us to tensors. Tensors and vectors are as intimately related,
as vectors and numbers. In fact, having studied the basics of tensor algebra, we will see that numbers, vectors, and tensors are conceptually

DOI: 10.1201/9781003620365-1

interconnected. We will be able to recognize that numbers are very *reduced* tensors[1]; it is said that numbers are tensors of *rank* zero. In a similar sense, vectors are incompletely reduced tensors; it is said that vectors are tensors of rank one. Therefore, the progression of the topics from numbers to tensors can be viewed as follows:

$$
\begin{array}{ccccc}
\text{Numbers} & \rightarrow & \text{Vectors} & \rightarrow & \text{Tensors.} \\
\text{Tensors}^{(0)} & \rightarrow & \text{Tensors}^{(1)} & \rightarrow & \text{Tensors}^{(2+)}.
\end{array}
$$

Here the superscript in parentheses indicates the rank of the tensor[2].

As we move from numbers to tensors, the level of abstraction increases. To a significant degree, the difficulty of understanding tensors is due to a high level of abstraction used in the definition of tensors as mathematical objects. Abstraction is the price we pay for more powerful and versatile tools. But more powerful tools are needed as scientists address more and more advanced problems.

The inventions of numbers, algebra, and then calculus were monumental breakthroughs. The transition from numbers to vectors and then to tensors is a more natural process that occurred rather quickly on the scale of the history of science.

1.1   WHO NEEDS TENSORS?

Today thousands of scientists – among them many physicists and mathematicians – use the methods of tensors. Tensor mathematics – the algebra and calculus of tensors – is a *tool*. It is a fitting tool for some problems, and not too fitting for others. This situation is quite analogous to *vector algebra and calculus*.

Tensor literacy will enrich you and will open doors to new problems and new methods of analysis. The following historical episode illustrates the point well.

In an October 1912 letter to a physicist Arnold Sommerfeld, Albert Einstein said the following:

---

[1]In a certain sense which will become clear after reading the book.

[2]Don't worry if the concept of *rank* seems unclear right now – it will be explained in due time.

Einstein on General Relativity

"I am now exclusively occupied with the problem of gravitation theory and hope, with the help of a local mathematician friend, to overcome all the difficulties. One thing is certain, however, that never in my life have I been quite so tormented. A great respect for mathematics has been instilled within me, the subtler aspects of which, in my stupidity, I regarded until now as a pure luxury. Against this problem [of gravitation] the original problem of the theory of relativity is child's play."

In the period from 1905 to 1916 Einstein was feverishly working on the General Theory of Relativity – the next best theory of gravity since Newton. The mathematics of General Relativity is based on the calculus of tensors, created by Italian mathematicians Ricci-Curbastro and Levi-Civita roughly a decade before Einstein started working on the problem of gravity[3].

To overcome the mathematical difficulties, Einstein used the help of his friend and former classmate Marcel Grossmann, who was an expert in tensor calculus and non-Euclidean geometry. The General Theory of Relativity was the first physical theory to use the power of tensors (in combination with profound physical insights) to achieve remarkable breakthroughs in understanding nature. Since then, the methods of tensor calculus and non-Euclidean geometries have been used in many physical theories and problems.

At the end of the book (Section 6.4 on page 147), a less dramatic example is given. The example describes a real-world situation when the understanding of vectors and tensors may lead to significant practical benefits.

1.2   NAIVE NOTION OF TENSORS

We may think of tensors as some kind of "super-numbers." In what sense tensors are numbers and what makes them "super?"

Similar to numbers, tensors can be added and subtracted. Also, tensors can be "scaled" by multiplying them by "normal" numbers like 2 or π.

---

[3]See *Einstein's Italian Mathematicians* by Judith R. Goodstein.

In contrast with numbers, tensors support a richer set of operations. Given two tensors, we can "kind-of-multiply" them to get either a simple number as the result (*scalar product*, Section 5.1), or we can get another tensor, somewhat "bigger" and more complex than the original two (*tensor product*, Section 5.8.)

Looking at the evolution of the concept of numbers, we can see the series of steps to higher levels of *generality, efficiency*, and *abstraction*: From natural numbers to whole numbers, to fractions, to real, and then to complex numbers. At each step new *mathematical objects* are introduced, and these new objects can be added and multiplied in a "usual way."

Tensors appeared as the result of quite the natural evolution of "number-like mathematical objects." Tensors extend the notion of numbers, all the way through vectors into a new and very powerful realm. If numbers are "bare quantities" and vectors are "quantities with direction" (e.g., velocity in physics), then tensors are "quantities with shape."

Tensors are naturally and closely connected with numbers and vectors. In fact, numbers and vectors *are tensors!*

Tensors Naively Defined

- Tensors are *mathematical objects* that cover and extend the concepts of numbers and vectors. As more powerful mathematical objects, tensors support many algebraic operations, including addition, subtraction, and scaling by a number.

- Tensors might be viewed as "quantities with shape," in analogy with vectors – "quantities with direction."

- Numbers and vectors represent the lowest "tiers" (called *ranks*) in the hierarchy of tensors.

- Finally, tensors generalize the idea of *linear functions* (see Subsection 2.3.3 on page 22 and also Section 4.2). Tensors of higher ranks can "act" on tensors of lower ranks in a very simple way, resembling familiar multiplication.

1.3   EXAMPLE DEFINITIONS

Now, what are tensors more rigorously? Can we give a short definition to this concept? Let us take a look at several examples and see whether they shed sufficient light. Although the definitions given below differ from each other, they convey *the same idea in different ways.*

The *Encyclopedia of Mathematics*[4] provides the following definition:

Tensors Definition 1

Tensor on a vector space V over a field k is an element t of the vector space
$$T^{p,q}(V) = (\otimes^p V) \otimes (\otimes^q V^*),$$
where $V^* = \text{Hom}(V, k)$ is the dual space of V.

To understand this definition, we first need to understand what *vector space* is, what *field* is, what *dual* means, and what is going on with superscripts and circles (e.g., in $\otimes^q$).

Wolfram Math World[5] provides another view of tensors:

Tensors Definition 2

An nth-rank tensor in m-dimensional space is a mathematical object that has n indices and m^n components and obeys certain transformation rules.

Here we encounter new concepts, such as: *rank of a tensor*, *m-dimensional space*, *indices*, *components*, and some kind of *transformation rules*. They all will be discussed later in the book.

Yet another definition can be found in the book *Encyclopedia of Mathematics*[6]

Tensors Definition 3

Just as a *vector* is a mathematical quantity that describes translations in two- or three-dimensional space, a tensor is a mathematical quantity used to describe general transformations in n-dimensional space. Precisely, if the locations of points in n-dimensional space are given in one coordinate system by $(x^1, x^2, \ldots, x^n)$ and in a transformed coordinate system by $(y^1, y^2, \ldots, y^n)$ (it is convenient to use superscripts rather than subscripts), then a "rank 1 contravariant tensor" is a quantity T, with single components, that transforms according to the rule:

$$T^i_{new} = \sum_{r=1}^{n} \frac{\partial y^i}{\partial x^r} T^r.$$

---

[4]https://encyclopediaofmath.org/wiki/Tensor_on_a_vector_space
[5]https://mathworld.wolfram.com/Tensor.html
[6]*Encyclopedia Of Mathematics*, James Tanton, Facts On File, Inc, 2005.

Again we face a wall of new concepts: *vectors*, *transformations*, *n-dimensional space*, *contravariant tensors*, and their *components*. Without clear understanding of these concepts, it is impossible to learn what tensors are.

The definitions given above are typical. They provide a good way to *end* the study of tensors, to summarize everything learned about them. However, they are not good starting points. It is better to arrive at the concept of tensor gradually, going from numbers, through vectors, to tensors. This path starts in the next chapters. But before we begin, a couple of preliminary remarks are needed.

1.4 DIAGRAMS

Sometimes to illustrate mathematical concepts and *relations between them*, we will use diagrams. Diagrams are helpful in highlighting some general features of *mathematical structures*.

Let us study an example, shown in Figure 1.1. A collection of cars in a parking lot – Figure 1.1(a) – can be schematically represented as a set of *points* Λ – Figure 1.1(b). Each point of the set Λ corresponds to a certain car in the parking lot. Apart from the set of cars Λ, Figure 1.1(b)

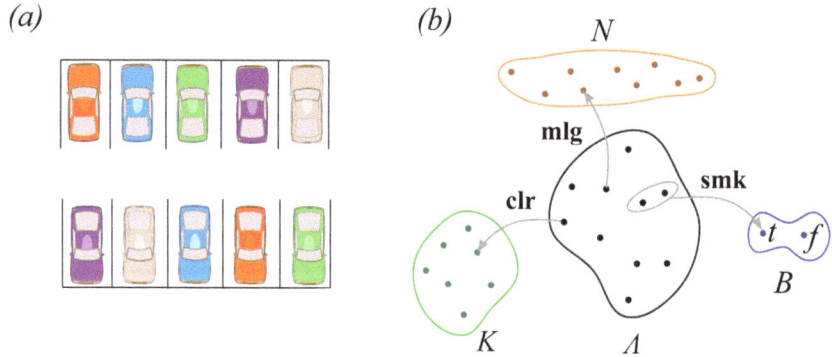

Figure 1.1 Diagrams are used to graphically represent sets of objects and relationships between them. (a) Cars in a parking lot as an example of a set. (b) Connections/relations between the set of cars and other sets, such as numbers, colors, and True/False sets. Arrows can connect (map) elements of one set with another. Such mappings may have names: **mlg** returns mileage for a given car, **clr** – color, and **smk** determines whether two cars are of the same make.

contains other sets, denoted as K, N, and B. These sets correspond to various values of cars' properties, such as color (set K), mileage (set N), and so on. The set B is a very important set – called a *Boolean set*. Boolean set has just two points, one labeled t for True, the other labeled f for False.

Reduced to a point of the set Λ, a car loses its individuality and those parameters that make it unique (make, color, mileage, etc.) This is remedied by using other sets, such as the set of various colors K or the set N of numbers that can represent mileage, and so on.

A particular property of a car-point can then be represented using an arrow that connects the car-point to another point in the relevant set. We say that such an arrow *maps* points of one set into another set. Figure 1.1(b) shows three possible maps: **mlg** gives the mileage for each car from the set Λ, **clr** gives the color for each car, and **smk** compares whether two cars have the same make. The names given to these mappings are in agreement with implicit mathematical convention to give functions a two or three letter name, like sin, log, or exp.

Exercise 1.1 *Extend the diagram from Figure 1.1(b), adding a set of different car makes (e.g., Ford, Toyota, Fiat, etc.). Come up with a mapping from this set into the Boolean set B.*

1.5 SCHEMATICS

To illustrate the concepts of functions, operators, their structures, and properties, we will be using schematics like the one shown in Figure 1.2.

A simple schematic element is represented as a box with inputs and outputs. A box can have a name (label) which describes what the

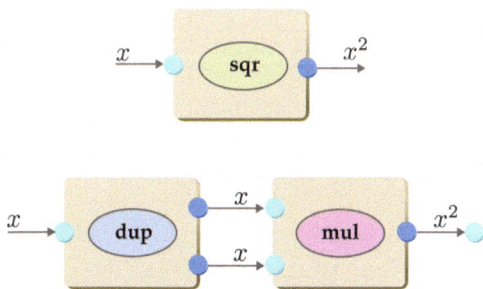

Figure 1.2 Schematics can be used to represent functions, operators, their compositions and structure.

function does to its input. The number of inputs and outputs can vary depending on the complexity of a function.

Various "boxes" can be combined (or *composed*) to create a more complex structure. Figure 1.2 shows how the outputs of a function **dup** (duplicate its input) are connected to the inputs of a function (**mul**) that multiplies its inputs. The result is a function that squares its input.

The placement of inputs on the left and outputs on the right sides of boxes is purely conventional. If preferred, equivalent schematics can be drawn when the inputs come from the right, like they do in more traditional notation of function application: $\sin \phi$ or $\exp x$.

1.6 SETS AND TUPLES

We will use many notational conventions in this book. Most of them will be typical for mathematics. For example, "+" denotes summation of two quantities, "=" means equality of two quantities, "$*$" – a multiplication, and the parentheses in "$3 * (x + y)$" are used to indicate the order in which operations should be performed (first add and then multiply).

At a certain point, we will be discussing "assemblies"[7] of quantities. The simplest example – all natural numbers:

$$1, 2, 3, 4, 5, \ldots, n, \ldots.$$

Or we can consider all letters of some alphabet:

a, b, c, d, e, f, g, h, i, j, k, l, m, n, o, p, q, r, s, t, u, v, w, x, y, z.

Both these examples can be formally considered as *sets* – an assembly of objects of similar kind. A set can be given a special name when it is often referred to. For example, the set of natural numbers is denoted as

$$\mathbb{N} = \{1, 2, 3, \ldots\}.$$

Note the use of curly braces – they indicate that we are talking about a set. Thus, for a set $\mathbb{S}$ with elements a, b, x, and y we write

$$\mathbb{S} = \{a, b, x, y\}.$$

The concept of a set is one of the most basic concepts in mathematics and the set notation is very standardized.

---

[7]Collections, ensembles, groups, and families – these terms already have reserved specific mathematical meanings, although they express similar ideas.

Another useful concept is called a *tuple*. A tuple is a series of quantities that are related in a certain way but can be of different kinds. It is better to study examples:

$$(3, \text{'a'}) - \text{couple (pair)},$$

$$(\vec{a}, \vec{b}, \theta) - \text{triple},$$

$$(x, 4, \vec{a}, \sin) - \text{quadruple}.$$

A series of n quantities is called n-tuple.

The most familiar use of tuples is the representation of coordinates of points in space. In two dimensional plane with Cartesian coordinate axes x and y we may have:

$$(x, y) = (2, 5) \quad - \text{couple (pair)}.$$

In three dimensional space:

$$(x, y, z) = (0, 1, -2) \quad - \text{triple}.$$

Notice the difference between the coordinate tuples and the examples given above: The tuples with coordinates contain only numbers, whereas, in the examples above, tuples contain quantities of various types (number and a letter for the pair, two vectors and a number for the triple, etc.). The point is that tuples *can* contain quantities of the same type, but in general, they do not. Another important difference between sets and tuples is the importance of the order of elements inside the tuple. Consider a set with two elements:

$$\mathbb{S} = \{0, 1\} = \{1, 0\}.$$

The order in which we write the elements does not matter. In contrast, the order is important for tuples:

$$(0, 1) \neq (1, 0).$$

This inequality becomes obvious if we interpret these pairs as coordinates of points in a plane. The first pair corresponds to the point on the x axis, while the second pair corresponds to the point on the y axis.

Finally, the tuples can be used to concisely describe *mathematical structures*. As a simple example, consider the set of whole numbers $\mathbb{Z} = \{0, \pm 1, \pm 2, \ldots\}$. We can do many arithmetic operations with these numbers, but let's focus only on the operation of addition. We can summarize our intention using a triple as follows:

$$(\mathbb{Z}, +, 0).$$

This expression says that we are studying a set of whole numbers $\mathbb{Z}$ equipped with a single operation "+". Moreover, we recognize that for this operation there is a special element 0 with "neutral" behavior:

$$n + 0 = 0 + n = n$$

for any number n from the set $\mathbb{Z}$. We will encounter more examples of this sort later in the book.

CHAPTER HIGHLIGHTS

- *The natural evolution of mathematical objects from numbers, through vectors, leads to tensors.*

- *Each successive tier of mathematical objects in the progression "numbers, vectors, tensors" is more abstract and more powerful.*

- *Numbers, vectors, and tensors are all conceptually interconnected.*

- *Just like the use of vectors opens up new methods for solving abstract and applied problems, the use of tensors opens up new, even more powerful, methods for solving problems in various domains of science.*

- *Diagrams and schematics are helpful for illustrating various mathematical relations and structures.*

- *A set is an assembly of objects of a similar kind. It is one of the most basic concepts in mathematics.*

- *A tuple is an ordered sequence of elements related to each other by a common context. The elements of a tuple can be of different kinds.*

CHAPTER 2

Numbers and Functions

N UMBERS ARE POWERFUL MATHEMATICAL objects. They are used in an endless list of problems that involve *quantities*. As mathematics and sciences progressed, natural numbers evolved into whole numbers, then into rational numbers and beyond.[1]

At a certain stage, physics required a mathematical tool to describe quantities with arbitrary directions. For example, the motion of a body involves velocity – a physical quantity describing how fast the body is moving and in what direction. Quite rapidly vectors led to tensors. Tensors had to be invented because there are many important problems where tensors are very natural. Examples will be given in the last chapter of the book.

At first glance, numbers, vectors, and tensors appear rather different. However, they have a lot in common. Tensors are "numbers on steroids" because all operations you can do with numbers, you can do with tensors – and even more.

Before we turn to tensors, we should familiarize ourselves with vectors. And before that, we must review the main concepts associated with numbers.

2.1 POWER OF ABSTRACTION

Mathematics is a remarkably effective and universal discipline, its methods and results can be applied in a wide range of fields. In part, the universality of mathematics stems from the *general* and *abstract* nature of mathematical concepts. Let us illustrate this using an example.

---

[1]A superb account of this process is given in the book *Number: The Language of Science* by Tobias Dantzig.

DOI: 10.1201/9781003620365-2

Figure 2.1 49 objects can be arranged in a square 7×7. 48 objects can be arranged as a rectangle of 6×8.

An astute farmer notices that 49 sacks of grains can be arranged in a square with each side having 7 sacks (see Figure 2.1). After one sack is used up, the remaining 48 sacks can be arranged as a rectangle of 6 by 8 sacks.

The farmer realizes that this curious fact has nothing to do with either grains or sacks. The same observation could be made about buckets, chairs, people, and so on. As the first step of generalization, the farmer states that 49 *objects* can be arranged as a 7 by 7 square, while 48 objects can be arranged as a 6 by 8 rectangle. The farmer also notices that $48 = 49 - 1$, whereas $6 = 7 - 1$ and $8 = 7 + 1$. She writes down the newly discovered relation as follows:

$$7 * 7 \text{ obj } - 1 \text{ obj } = (7 - 1) * (7 + 1) \text{ obj},$$

where *obj* is the denotation of *any* object.

As the grain is used up, the farmer discovers two more relations:

$$6 * 6 \text{ obj } - 1 \text{ obj } = (6 - 1) * (6 + 1) \text{ obj}$$

and

$$5 * 5 \text{ obj } - 1 \text{ obj } = (5 - 1) * (5 + 1) \text{ obj}.$$

At this point the farmer makes an educated guess, stating that a more general relation must exist:

$$n * n - 1 = (n - 1) * (n + 1). \tag{2.1}$$

In the last expression, the reference to objects is dropped and the expression is written purely in terms of *numbers*.

A deeper analysis reveals that the relation given by (2.1) exists for *any quantities* that obey the usual rules of addition and multiplication.

This includes rational numbers, real numbers, complex numbers (see Section 6.2), and even operators[2]! The relation

$$x^2 - 1 = (x - 1)(x + 1) \tag{2.2}$$

holds true because of the way we define *rules for manipulation* – addition and multiplication in this case – of number-like objects, regardless of what those number-like objects represent in a particular problem.

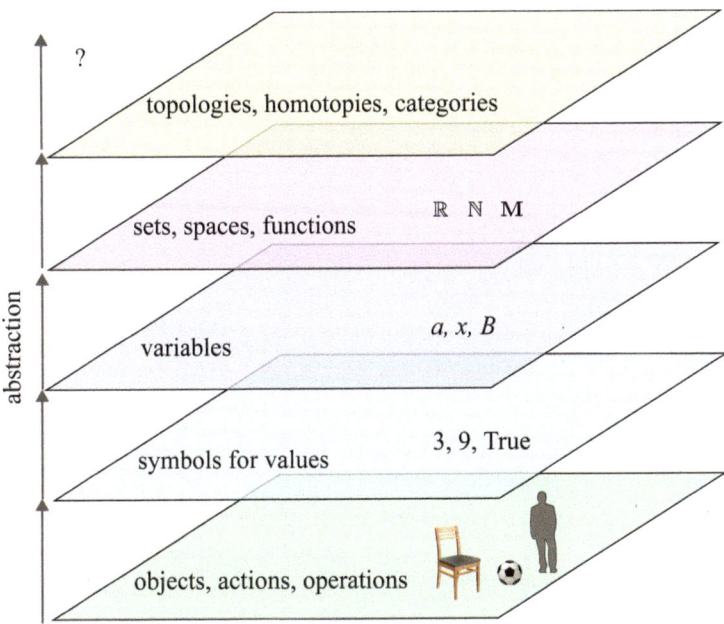

Figure 2.2 Mathematical thinking has many levels of abstractions. Going to higher levels of abstractions results in higher efficiency and more powerful ideas and tools. It also becomes more challenging to understand and master more advanced concepts and methods.

The path from "sacks of grain" to a variable x is the path from concrete, specific objects to *abstract* entities that are the product of *creative imagination*. This path to higher levels of abstraction is illustrated in Figure 2.2. As we move to higher levels of abstraction, our mathematical tools become more powerful and more universally applicable: From everyday arithmetic to economics, to general relativity, and quantum gravity.

[2]For now think of operators as functions that act on non-numeric arguments.

Using more abstract mathematical objects requires serious mental effort. To reach the highest levels one needs to do mathematics professionally. However, any profession can benefit from *some* level of abstraction, and to understand vectors and tensors, we must go beyond the usual high-school level. One of the goals of this book is to help readers build tolerance and appreciation of more abstract aspects of mathematics.

2.2 TERMINOLOGY BARRIER

Every high school student has a working knowledge of bilinear operators over associative and commutative fields, but hardly any of those students are aware of this fact. We refer here to the ability to simply add and multiply numbers. This demonstrates that even familiar and basic notions may look complicated when "dressed in unfamiliar clothing."

When we learn new mathematical concepts, especially at a higher level of abstraction, we often encounter what might be called a *terminology barrier*: a concept seems more difficult if it is formulated in a new language, without sufficient connections to already familiar concepts, and without clear examples of how the concept can be applied.

The new terminology is unavoidable when learning new concepts. There will be a number of new mathematical concepts and definitions introduced in this book. To lower the terminology barrier, every new mathematical concept will be illustrated with examples and connections to already familiar concepts. Additionally, it is recommended to do the following exercise every time a new concept with unusual terminology is introduced:

Dealing With New Concepts

- Take a critical look at a new name and notation.

- Think whether the new name or notation looks like something you know. Is the resemblance helpful or misleading?

- Be creative and try to come up with your own notation or word to describe the new concept.

  Remember: Symbols and names are not essential. What is important is the set of *relations* of a new concept to other concepts. The relations show how the concept fits and functions within the larger framework.

Demonstrations of this approach can be found in the rest of the book.

2.3 ALGEBRA OF NUMBERS

We will start with the "familiar" numbers:

$$0, \pm 1, \pm 2, \ldots, \pm n, \ldots$$

This endless collection, considered as one entity, is called the *set of whole numbers*; it is denoted as $\mathbb{Z}$.

The set $\mathbb{Z}$ is not a formless heap of elements. On the contrary, it has a rich *structure*. The structure of any set, including the set of whole numbers, is revealed through various *relations* between all or some of its elements. Here are several examples:

- 1 and −1 are related, and so are 2 and −2, and generally, n and $-n$.

- Relation exists between 1 and 2, 2 and 4, 3 and 6, and generally, between n and $2n$.

- The pairs $(1, 2)$, $(2, 3)$, $(3, 4)$, and so on, illustrate an important relation of order that exists in $\mathbb{Z}$.

- The pairs $(2, 1)$, $(3, 1)$, $(4, 2)$, $(5, 1)$, $(6, 3)$, and similar ones unite a number and its greatest divisor is not equal to the number itself.

The list of relations can be continued indefinitely, but the general idea of relations should be clear. Such number-to-number relations can be schematically represented as boxes with inputs and outputs, as shown in Figure 2.3. A few important relations are given descriptive names: **neg**, **dbl**, **suc**, and **gsd** are examples from Figure 2.3; they correspond to the negation of a number, doubling of a number, finding the successor of a given number, and finding the greatest divisor smaller than the number itself.

2.3.1 Functions

What we have just described is the idea of a *function*, or numeric function of a single *argument*, to be precise. A function of a single argument connects every *argument* (input) to a certain *value* (output), establishing a *relation* between a pair of elements.

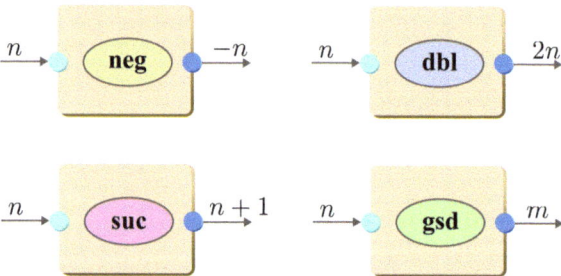

Figure 2.3 Relations between elements of a set can be schematically represented using boxes with inputs and outputs. Here the relations between numbers are given descriptive names: **neg** is negation, **dbl** is doubling, **suc** is getting the successive number, **gsd** is the greatest divisor of a number smaller than the number itself.

Another view on relations is illustrated in Figure 2.4. Relations between elements can be "elevated" to the level of sets and depicted as arrows connecting one set (*domain*) to another set (*range*). Symbolically we can write:

$$\mathbb{Z} \xrightarrow{\text{dbl}} \mathbb{E},$$

where $\mathbb{E}$ denotes the set of all even numbers, **dbl** is the name of the function that doubles its argument.

The relations of the type "one number to one number" – considered above – can be generalized to "several numbers to one number," or "one number to several numbers," or even "several numbers to several numbers." The schematic representation of such relations is given in Figure 2.5.

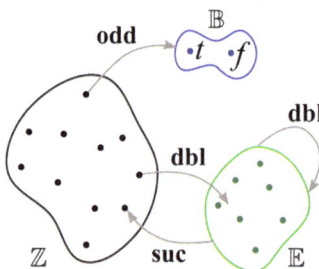

Figure 2.4 Relations can be viewed on the level of sets. A function maps (connects) one set with another in a meaningful way. For example, **dbl** maps every integer from $\mathbb{Z}$ into the set of even numbers $\mathbb{E}$.

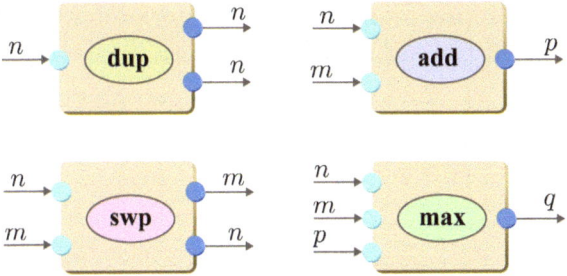

Figure 2.5 Relations of several elements to several elements: **dup** duplicates its input, **add** calculates the sum, **swp** swaps the order of the arguments, **max** returns the maximum of three input numbers.

Exercise 2.1 *Think how you would represent the generalized relations of the types given in Figure 2.5 at the level of sets. What kind of diagrams would you draw?*

2.3.1.1 Binary Functions

An important function of two variables is the familiar **add** relation:

$$\mathbf{add}\, n\, m = n + m.$$

The right-hand side of this equality is just another way of writing the expression involving function with two input arguments. It is a special case of a more general rule, which can be written as follows:

$$f\, n\, m = n \odot m.$$

On the left, we have a *prefix* notation, where a function f is *applied* to two arguments[3]. On the right, we have an *infix* notation and a special symbol placed between the first and the second argument. Several familiar examples are

- **mul** $n\, m = n * m$ – multiplication.

- **pow** $n\, m = n^{\wedge}m$ – power.

- **sub** $n\, m = n - m$ – subtraction.

- **div** $n\, m = n/m$ – division.

---

[3]The use of parentheses around the arguments will be discussed shortly.

The functions **mul**, **pow**, **sub**, and **div** are all examples of *binary* functions – functions of two arguments.

Exercise 2.2 *Think of your own example of a binary function. Create its infix variant.*

2.3.1.2 Unary Functions

Functions of single argument are called *unary* functions. Various notations exist to describe the action of unary functions on their arguments. The most often used notation looks like this:

$$f(x) = y.$$

In this book, we will use another notation to express that the function f is *applied* to its argument:

$$f\,x = y.$$

The parentheses will only be used to write an argument with structure:

$$f\,(x + y * z) = v.$$

Without the parentheses, the left-hand side of the last expression would be understood as the sum of two terms:

$$f\,x + y * z = (f\,x) + (y * z).$$

It is said that function application has the highest *precedence* in an expression.

Parentheses Or Not?

Writing simple arguments without parentheses is a standard practice in trigonometry. For example:

$$\sin \alpha + \sin \beta = 2 \sin \frac{\alpha + \beta}{2} \cos \frac{\alpha - \beta}{2}.$$

One can find many similar examples in books and papers. Consider, for instance, an iconic book *Mathematical Methods For Physicists* by George B.

Arfken and Hans J. Weber[4]. There we find:

$$(6.12) \quad \cos n\theta + i \sin n\theta = (\cos \theta + i \sin \theta)^n, \text{ or}$$

$$(6.13a) \quad \ln z = \ln r + i\theta, \text{ or}$$

$$(10.80b) \quad \text{erf}\, z = \pi^{-1/2} \gamma \left(\frac{1}{2}, z^2 \right).$$

For many the habit of using parentheses is very strong and a serious effort is required to accept the alternative. Surprisingly, however, when one starts working with *operators* – which are essentially different kinds of functions – the resistance disappears and it becomes a universal practice **not to put parentheses** around simple arguments. Here are some examples from the same book:

$$(1.117) \quad \vec{F} = -\vec{\nabla}\phi, \qquad \vec{\nabla} - \text{is an operator,}$$

$$(3.27) \quad x' = A\,x, \qquad A - \text{is an operator}$$

The moral is this: We should not let our habits thwart learning new and helpful things.

Sometimes we will write the relation between numbers in the following way:

$$x \xrightarrow{f} y \quad \text{or} \quad \mathbb{Z} \xrightarrow{f} \mathbb{E}.$$

The first variant emphasizes the relation between elements of a set, while the second expresses the same relation on the level of sets (whole numbers and even numbers, in this example). In both cases, f denotes the name of the function/relation.

Arity of a Function

Arity is the characteristic of a function describing the number of its arguments: unary (1 argument), binary (2 arguments), and ternary (3 arguments). The function of n arguments is called n-ary. Unary and binary are the most often used terms.

---

[4] Academic Press, Fourth edition, 1995.

2.3.2 Indexed Objects

Unary functions from the set of whole numbers $\mathbb{Z}$ provide a good introduction to *index notation* – an important tool in tensor mathematics.

Let us start with the simplest case of *counting* elements of some set $\mathbb{S}$. When counting, we label each element of the set with a unique number, thus establishing a *relation* between a number and an element of a given set $\mathbb{S}$. Put differently, we define a function from the set of whole numbers (if we allow negative numbers to be used as labels) into the set $\mathbb{S}$, as shown in Figure 2.6.

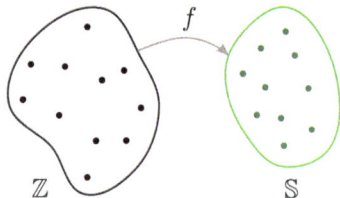

Figure 2.6 Counting elements of a set $\mathbb{S}$ creates a function from the set of whole numbers $\mathbb{Z}$ into the set $\mathbb{S}$.

We can express counting symbolically as follows:

$$i \xrightarrow{\ f\ } s_i,$$

where s_i is the element of the set $\mathbb{S}$ labeled with the number i (i is for *index*).

In many mathematical and physical problems, the set being counted is also made of numbers; in other words, the function f maps whole numbers into other numbers. Consider the function

$$f\,i = 2i + 1.$$

If we denote the output value of the function f as v, then all values can be indexed as

$$v_i = f\,i = 2i + 1.$$

Using subscripts for indices is very common, it avoids confusion with the power notation:

$$v_1 \neq v^1, v_2 \neq v^2, \ldots, v_n \neq v^n.$$

When the confusion is not possible or easy to avoid, placing indices as superscripts might be helpful (see Section 5.8 for details).

Continious Index

What if extend the idea of counting and allow "the label" to be a continuous variable? For example:

$$v_x = f\,x = 2x + 1.$$

This is just another way of writing an unary function f! We thus arrive at the idea of using a subscript to represent the argument of a function. Here is an illustration from the Arfken and Weber book *Mathematical Methods For Physicists*:

$$(10.67) \quad B_x\,(p,q) = \int_0^x t^{p-1}(1-t)^{q-1}dt.$$

Don't be intimidated by the right-hand side of this equation, notice that the expression of the left describes a *ternary* function, with the arguments x, p, and q.

2.3.2.1 *Multi-Index Objects*

It is possible, and often convenient, to label objects with several indices. This results in *multi-index* objects. Let us start with two indices and look at several examples, schematically represented in Figure 2.7.

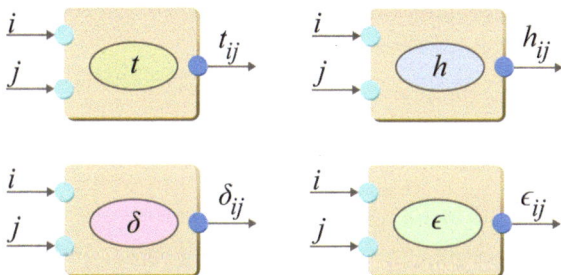

Figure 2.7 Mapping two whole numbers to another number can be described using indexed objects with two indices. Schematically they are represented as boxes with two inputs and one output. See the text for the definition of these particular objects.

First, if we denote as t_{ij} the approximate number of hours elapsed since the beginning of the year upto the i-th day of the j-th month, then we will have

$$t_{ij} = 24 * \left[30(j-1) + i - 1\right].$$

Second, denote h_{ij} the values of the following function:

$$h\,i\,j = \begin{cases} 0 & \text{if } j \leq i. \\ 1 & \text{if } j > i. \end{cases}$$

Third, denote δ_{ij} the values of the function

$$\delta\,i\,j = \begin{cases} 1 & \text{if } i = j. \\ 0 & \text{if } i \neq j. \end{cases}$$

Finally, denote ϵ_{ij} the values of the function

$$\epsilon\,i\,j = \begin{cases} 0 & \text{if } i = j. \\ 1 & \text{if } i < j. \\ -1 & \text{if } i > j. \end{cases}$$

The object δ is very useful in mathematics. It is called Kronecker delta, after German mathematician Leopold Kronecker. With the help of Kronecker's δ many mathematical expressions can be written compactly and many manipulations can be carried out efficiently, as we will soon see.

Objects with three or more indices appear in certain problems. We will define and discuss them when they become relevant for the study of tensors.

2.3.3 Linear Functions

We will now encounter for the first time the deceptively simple concept of *linear functions*. Tensors represent a powerful extension of this idea.

Consider the following two functions:

$$\textbf{dbl}\,n = 2n \quad \text{and} \quad \textbf{sqr}\,n = n^2.$$

The first function has an important property which the second one is lacking. Namely, we have

$$\textbf{dbl}\,(n + m) = (\textbf{dbl}\,n) + (\textbf{dbl}\,m)$$

and

$$\textbf{dbl}\,(a * n) = a * (\textbf{dbl}\,n).$$

Compare this to the function **sqr**:

$$\mathbf{sqr}\,(n + m) = (\mathbf{sqr}\,n) + (\mathbf{sqr}\,m) + 2nm \neq (\mathbf{sqr}\,n) + (\mathbf{sqr}\,m),$$

$$\mathbf{sqr}\,(a * n) = (\mathbf{sqr}\,a) * (\mathbf{sqr}\,n) \neq a(\mathbf{sqr}\,n).$$

The function **dbl** is an example of a *linear function*, while **sqr** is not.

Linear Functions

A *linear function* f is any function that satisfies the two simple conditions, called *linearity* conditions:

$$f\,(n + m) = (f\,n) + (f\,m)$$

and

$$f\,(a * n) = a * (f\,n).$$

These conditions severely limit what a linear function can do.

In the vast family of functions, some are simpler than others. The simplicity of a function is reflected in the conditions it satisfies. The more conditions we place on a function – the simpler it becomes. An arbitrary function f is not limited by any constraints. Linear functions are among the simplest. Other categories of functions, together with their constraints, are given in the table below:

Category	Condition	Example
Symmetric	$f\,(-x) = f\,x$	x^2
Anti-symmetric	$f\,(-x) = -(f\,x)$	$1/x^3$
Periodic	$f\,(x + a) = f\,x$	$\cos\,(2\pi x/a)$
Affine	$f\,(x + y) = (f\,x) + (f\,y)$	$4x + 7f$
Homogeneous	$f\,(a * x) = a^n(f\,x)$	$3x^n$
Linear	$f\,(ax + by) = a(f\,x) + b(f\,y)$	$x/2$

As a side-note, linear functions are both *affine* and *homogeneous* for $n = 1$.

Exercise 2.3 *Show that for any linear function we must have*

$$f\,0 = 0.$$

Then show that

$$f\,n = n(f\,1).$$

The last property demonstrates that the action of a linear function on any number is completely determined by a single value $f_1 = (f\,1)$ – its action on the number one.

Unary linear functions acting on numbers seem too simple to be of importance. They are equivalent to multiplication by a pre-defined value. Indeed, the action of a linear function f on any number reduces to the multiplication of that number by a fixed value $f_1 = f\,1$.

Any example of a linear function over numbers will appear trivial:

$$f\,x = 3\,x.$$

However, linear functions over vectors (and tensors) turn out to be very useful. The rest of the book aims to demonstrate this.

Example
Linear Functions

What is more important is the *concept of linearity*, as it can be more fruitfully applied to other mathematical objects. We will encounter linear functions in different contexts when we talk about *operators* – functions defined on vectors. Linearity in that case will not be trivial, but still important and easy to understand.

Exercise 2.4 *Draw a schematic representation of the linearity properties.*

2.3.3.1 Multilinear Functions

The idea of linearity can be generalized. A function with two or more arguments can satisfy linearity properties for each argument and this way become a *multilinear function*.

Let us consider a familiar function

$$\mathbf{mul}\ n\,m = n * m.$$

The distributivity of multiplication implies that

$$\mathbf{mul}\ (n + p)\,m = (\mathbf{mul}\ n\,m) + (\mathbf{mul}\ p\,m),$$

and

$$\mathbf{mul}\ n\,(m + q) = (\mathbf{mul}\ n\,m) + (\mathbf{mul}\ n\,q).$$

The second linearity property is also satisfied for each argument:

$$\mathbf{mul}\ (an)\ m = a(\mathbf{mul}\ n\ m),$$

and

$$\mathbf{mul}\ n\ (am) = a(\mathbf{mul}\ n\ m).$$

Multiplication is a *bilinear function*. The ideas of linearity, bilinearity, and multilinearity are important for a better understanding of tensors.

Now that we are familiar with the concept of multilinear functions, we can more fully appreciate other definitions of tensors:

Wikipedia on Tensors

The first paragraph on tensors in Wikipedia states[5]:

"*In mathematics, a tensor is an algebraic object that describes a multilinear relationship between sets of algebraic objects related to a vector space.*"

Here the term *multilinear* means exactly what we have just learned. The meaning of the rest of the phrase will become more clear as we progress through the book.

2.3.4 Function Composition

New functions can be built from already available functions. Several "natural" and simple methods exist for this. Let's study these methods using, for the purpose of illustration, two trigonometric functions:

$$f\ x = \cos\ x, \qquad g\ x = \sin\ x.$$

First, we must understand that it is permissible to talk about functions without mentioning their arguments. For example, we are allowed to write

$$f = \cos, \qquad g = \sin.$$

This way we emphasize our attention to functions as mathematical objects in their own right. This is similar to how we use numbers: We write just 5, not 5 *of something*, even though that *something* is always implicitly present.

---

[5] As of October 9, 2023.

Argument-Free Notation

The practice of manipulating functions without arguments (i.e., writing their names only) is called *argument-free* or *point-free* notation. Later, when discussing linear operators, it will become very helpful.

Now we can define meaningful operations on functions, analogous to arithmetic operations on numbers. First, we can add functions:

$$u = f + g.$$

The application of this new function u on any argument results in

$$u\,x = (f\,x) + (g\,x).$$

Adding Functions

Having defined an addition of two functions, we can write expressions like

$$f = \cos + \sin \ \ \text{or} \ \ h = \log + \exp.$$

It is important to remember that in these expressions we *do not add numbers*, we add functions – completely different mathematical objects.

In a similar way we can define the multiplication of functions:

$$v = f * g \quad \rightarrow \quad v\,x = (f\,x) * (g\,x).$$

Thus, we gave a meaning to the following expressions:

$$\cos + \sin, \quad \cos * \sin.$$

Lastly, unlike numbers, functions can be *composed* or sequenced. For example, we can write

$$h\,x = \cos(\sin x) \ \ \text{or} \ \ w\,x = \sin(\cos x).$$

Note that $h\,x \neq w\,x$.

More generally, and dropping the argument x, we can compose functions f and g in two different ways:

$$h = g \circ f \ \ \text{or} \ \ w = f \circ g.$$

Here we used infix notation and a conventional denotation for a composition operation "∘."

The schematic representation of the idea of function composition is given in Figure 2.8. Note how the order of the boxes differs from the order of the functions in the expression for composition. This is the consequence of placing the input to the boxes on the left and reading the schematic from left to right.

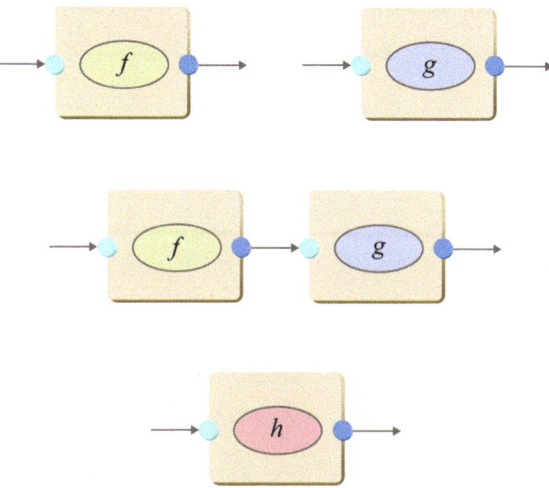

Figure 2.8 Functions can be *composed* to create a new function. In composition the output of the first function is "connected" to the input of the second.

In contrast to addition and multiplication, function composition does not possess *commutativity*. That is, in general we have

$$g \circ f \neq f \circ g.$$

For example,

$$(\cos \circ \sin) \, 0 = 1 \quad (\sin \circ \cos) \, 0 = \sin 1 \neq 1.$$

Sometimes, however, commutativity is satisfied, as we will later see in discussions of linear operators (see the end of Section 4.1.)

With the definitions given above, we can see the meaning of complex expressions like this one:

$$h = (\mathbf{sqr} \circ \cos) + (\mathbf{dbl} * \sin).$$

When applied to an argument, this function returns

$$h\, x = \cos^2 x + 2x \sin x.$$

At this point, the benefit of such abstract exercises is not evident. However, they open a door to understanding how functions and operators[6] can be manipulated and how their properties can be discussed in a manner analogous to more primitive mathematical objects, such as numbers. For example, we will be able to understand how a function may *become an argument to another function*, as in

$$\mathbf{sqr}\ \cos = \cos * \cos .$$

With this insight, the famous Euler's identity

$$e^{i\pi} + 1 = 0$$

will have a simple and clear interpretation (see formula (6.5) in Section 6.2 on page 136).

2.3.5 Partial Application

Any binary function can be turned into an unary function if we fix one of its arguments. Schematic representation of functions provides an especially clear view of this procedure, shown in Figure 2.9. When a

Figure 2.9 Functions with several arguments can be reduced to functions with even fewer arguments by *partially applying* – providing arguments to some, but not all of its inputs. In this example, partially applying the binary function **mul** we get a unary function **dbl** which doubles its argument.

function is applied to some of its argument – but not to all – it is said to be *partially applied*. The applied arguments are considered fixed, and the remaining arguments are considered as varying. Partial application reduces the arity of a function.

---

[6]Operators are defined later in Chapter 4.

We will encounter partial application later in the book when discussing two kinds of vectors: covariant and contravariant vectors (see subsection 5.4.1).

Partial Application

Partial application is an under-appreciated concept. It is uncommon in mathematics and physics, but more likely to be encountered it in the area of computation. The situation is slowly changing and today one can encounter examples of partial application, for instance, in modern papers on quantum physics:

"Assuming time-translation invariance, let us denote the transport function as $T(t_1 - t_0, \bullet)$ where the free slot is for the hidden variables."

This is a quote from the paper *The wave function as a true ensemble* by Jonte Hance and Sabine Hossenfelder[7]. What the authors did here is to apply the binary function T to the first argument with the value $t_1 - t_0$, leaving the second argument as the "free slot."

2.3.6 Function Representations

We explored the idea of functions as relations between elements of sets. So far we mentioned three different ways to represent this idea:

1. *Schematically*: using boxes with inputs and outputs.

2. *Diagrammatically*: using arrows connecting elements of a set of different sets.

3. *Symbolically*: using mathematical formulas made from variables and operations.

For completeness, let's take a look at two more ways to represent functions.

2.3.6.1 *Functions as Graphs*

Numeric functions with single input and single output can be plotted as curves in a plane. Figure 2.10 shows two curves: parabola for the **sqr** function, and sinusoidal curve for sin. Such graphs visually demonstrate

---

[7]Hance Jonte R. and Hossenfelder Sabine, 2022, *The wave function as a true ensemble*, Proc. R. Soc. A.47820210705, http://doi.org/10.1098/rspa.2021.0705

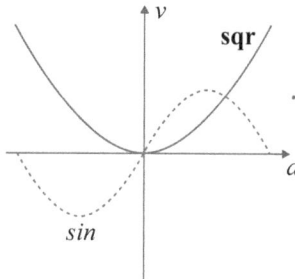

Figure 2.10 Simple functions (with one input argument a and one output value v) can be represented as a curve in a plane with two axes: a for argument, and v for value. Two functions are plotted here: $v = \mathbf{sqr}\, a = a^2$ and $v = \sin a$.

the relation between argument a and value $v = f\, a$ for a given function f. This is a powerful method for studying functions and their properties. For instance, we can see the presence and locations of maxima, minima, points of zero-values, regions of positive and negative values, and so on.

Representing functions as curves has its limits. Functions with several inputs or several outputs often present significant challenge for clear and informative plotting. In such cases, we can resort to other representations of functions, for example using symbolic mathematical expressions.

2.3.6.2 Functions as Tables

Functions may also be represented as tables. This method works best for functions whose arguments take on a discrete and limited set of values. For example, if we record the temperature at noon every day during the year, we will experimentally establish a relation between the day number and the reading of the thermometer, as summarized in the Table 2.1.

TABLE 2.1 Functions can be represented as tables that relate arguments to values in a series of entries. Here, for every value of the argument d from the first row, the value of t is given in the second row.

d	1	2	3	...	n
t	t_1	t_2	t_3	...	t_n

Tabular form is also convenient in practical applications when the precision of the function's values is not critical and the function can be approximated by a series of values at a limited number of points.

We explored five different ways to talk about functions. The main point of this exploration can be formulated in the following way:

Concept and Its Representation

The mathematical concept of a function as the relation between elements of sets allows several representations. Each representation has its merits and each representation highlights certain aspects of the concept.

None of the representation is definitive, none fully conveys what functions are. Different representations show *different aspects* of *the same idea*: They are different doors into the same room.

It is important to understand the *relation* between mathematical concepts and their various *representations*. This theme will keep coming up as we learn more about numbers, vectors, and tensors.

2.3.7 Numbers

Modern understanding of numbers differs greatly from how numbers were viewed in the past. From puzzling zero and negative numbers to tricky fractions and abstract real numbers, the family of mathematical objects called numbers grew to include natural numbers, whole numbers, rationals, real, and complex numbers.

Now, what defines a number? What properties does a mathematical object have to satisfy to be accepted into the "number family"?

To be a number an object must "behave like a number": We must know how to add and multiply it with other numbers to get a numeric answer as the result. In other words, numbers form a *set* equipped with two binary functions – addition and multiplication. These functions are simple and satisfy several requirements:

$$x + y = y + x \text{ (commutativity)}.$$

$$x * y = y * x \text{ (commutativity)}.$$

$$z * (x + y) = z * x + z * y \text{ (distributivity)}.$$

2.3.7.1 Neutral Elements

There are two special numbers: 0 for addition and 1 for multiplication. Their special character is expressed by the following equations:

$$0 + x = x \text{ for all } x,$$

and

$$1 * x = x \text{ for all } x.$$

The number 0 is the neutral element with respect to addition, and the number 1 is the neutral element with respect to multiplication.

2.3.7.2 Inverse Elements

Every number (except zero) has two "anti-numbers" defined in a natural way. In addition, every number x has its "anti-number" $\bar{x}$ such that

$$x + \bar{x} = 0.$$

Similarly for multiplication, for every number x there exists its "anti-number":

$$x * \tilde{x} = 1.$$

Here we deliberately deviated from the conventional notation: For the inverse of x with respect to addition we would have to write $(-x)$, and for the inverse with respect to multiplication: $(1/x)$ or (x^{-1}).

In the following chapters, we will encounter new mathematical objects, such as vectors and tensors. Like numbers, they will be elements of some sets and, like numbers, their pairs can be added and multiplied. Like numbers, they may have neutral and inverse elements. And yet they will differ from numbers in several important ways.

To better understand the distinct nature of vectors and tensors, it is important to be able to look at the *structure* of their corresponding sets.

2.3.8 Algebraic Structure

Numbers, as well as vectors and tensors, form *structured sets* in the sense that there are *relations* between their elements. For numbers, two important relations are the familiar operations of addition and multiplication. In mathematics, any set structured this way is called a *field*. This is the meaning of the word "field" in the *Tensor Definition 1* on page 5.

A set of elements, together with operations on the elements form what mathematicians call an *algebraic structure* or simply *algebra*. Numbers form an algebraic structure, and so do vectors and tensors.

Algebra of Functions

Think of how functions, such as sin, cos, f, and so on, form an algebra when the operation of *composition* is given. Such an algebra is the subject of *category theory*.

The essential differences between numbers and, for example, vectors, are *not* how they are represented symbolically or graphically, but what is the *structure* of their sets, what are the *operations* and their *properties*.

The set of real numbers $\mathbb{R}$ equipped with usual operations of addition and multiplication forms an algebraic structure called *field* and it is denoted using the tuple notation (see Section 1.6) as follows

$$(\mathbb{R}, +, 0, *, 1).$$

Here we list the set, the operations with its elements, and the special elements of those operations.

Vectors[8] (introduced in Chapter 3) also form a set. The elements of this set can be added to yield another vector. If we denote a pair of vectors as $\vec{a}$ and $\vec{b}$, and the operation of their addition as "$\vec{+}$," then their addition can be described by a vectorial equation:

$$\vec{a} \mathbin{\vec{+}} \vec{b} = \vec{c}.$$

Note that a different symbol "$\vec{+}$" is used to represent the addition of two vectors! We have to remember that despite being similar, number addition and vector addition are *different operations* after all! Indeed, we are operating on elements of different natures and our notation should reflect that. However, nearly always we can be less pedantic and use the traditional symbol for addition:

$$\vec{a} + \vec{b} = \vec{c}.$$

---

[8]Although we did not study vectors yet, some of the points made here might be helpful. It is recommended to come back to this section after reading Chapter 3.

Numbers can be defined and used without vectors, while vectors *can not* be defined or used without numbers! A simple equation

$$\vec{a} + \vec{a} = 2\vec{a}$$

requires not only a number but also a well defined operation of multiplication of a vector and a number:

$$2 \overset{\rightarrow}{*} \vec{a} = \vec{d}.$$

Again, the arrow on top of the asterisk "$\overset{\rightarrow}{*}$" can usually be dropped, as long as we remember that the multiplication of two numbers is *not the same* operation as the multiplication of a number and a vector.

Thus, from the more abstract standpoint, vectors form an algebraic structure that can be denoted as follows:

$$(\mathbb{V}, \overset{\rightarrow}{+}, \vec{0}, \overset{\rightarrow}{*}, 1).$$

Here the capital $\mathbb{V}$ represents *all possible* vectors – *vector space* in mathematical jargon; $\vec{0}$ is a special neutral element with respect to the addition of vectors $(\overset{\rightarrow}{+})$, analogous to the number zero; $(\overset{\rightarrow}{*})$ denotes the operation of multiplying a vector by a number; and 1 is the familiar number, the same neutral element with respect to numeric multiplication.

Numbers and vectors form different algebraic structures, at least for the following two reasons: 1) Numeric neutral element 1 has no counterpart among vectors (no multiplicative vector $\vec{1}$); 2) Multiplication of two numbers yields a number, whereas in the definition of vectors, there is no operation of multiplication of two vectors that would result in a new vector – we only multiply a number and a vector[9].

Returning to the Tensor Definition 1 on page 5, we now should be able to understand the meaning of the phrase: *Tensor on a vector space V over a field k*... Here tensors are defined based on the idea of vectors, which, in turn, require the notion of numbers (or other number-like mathematical objects that form *field*). We should expect tensors to form an algebraic structure different from both numbers and vectors. That structure will be more clear once we have learned the concepts of vectors and linear operations on them.

---

[9]Vector multiplication is a tricky operation, especially for vectors with dimensions higher than three.

2.4 POWER OF NOTATION

The power of mathematics that comes from abstraction is amplified by a good notation. Notation is a tool which helps express ideas clearly and manipulate them efficiently. Notation plays an especially important role in mathematics.

Learning mathematics requires learning new and sometimes non-intuitive notation. Effort is needed to learn new notation, and, more importantly, it is necessary to learn to see beyond symbols in order to recognize *concepts* and *relations* between them. One of the goals of this book is to help learn new notation, and make sense of unfamiliar, and at times strange-looking, symbols used in mathematics.

Having finished this book, the reader should be able to appreciate the expressiveness of formulas like this one:

$$T^{p,q}(V) = (\otimes^p V) \otimes (\otimes^q V^*).$$

This expression is used in one of the definitions of tensors given in the introduction (see Section 1.3 in Chapter 1).

2.4.1 Polynomial Example

As an example, let us consider a simple mathematical function given by the following cubic polynomial:

$$P_3\, x = 2x^3 - 2x^2 - 4x. \tag{2.3}$$

Imagine we would like to understand the structure and properties of this polynomial. We can use various approaches and examine the function from different viewpoints. For example, we can examine the schematic representation of the polynomial shown in Figure 2.11. The schematic, built from more basic "building blocks," reveals a moderately complicated structure of the polynomial. It is definitely not as compact as the symbolic expression (2.3).

Algebraic manipulations lead to an alternative way of writing the same polynomial:

$$P_3\, x = 2x(x + 1)(x - 2). \tag{2.4}$$

The schematic representation of (2.4) is shown in Figure 2.12. Although the schematic of the factorized expression looks simpler than for the non-factorized, it still looks unwieldy compared to the symbolic form of (2.4).

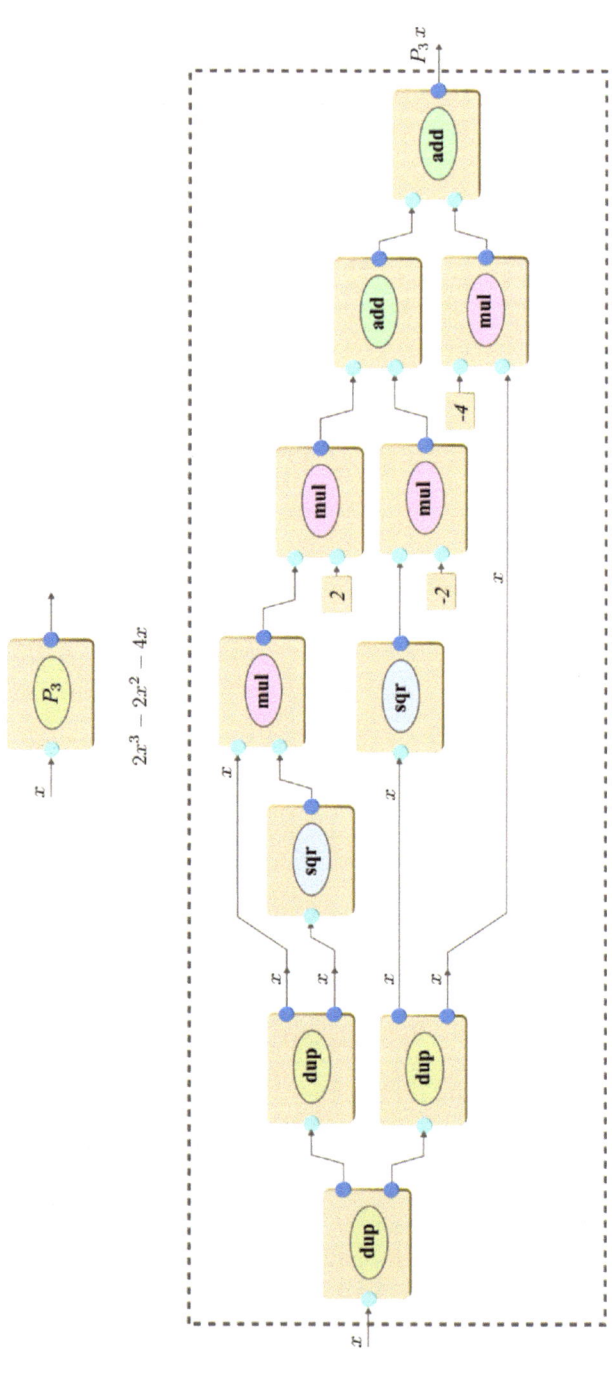

Figure 2.11   Schematic representation of the structure of a cubic polynomial, written symbolically as $2x^3 - 2x^2 - 4x$.

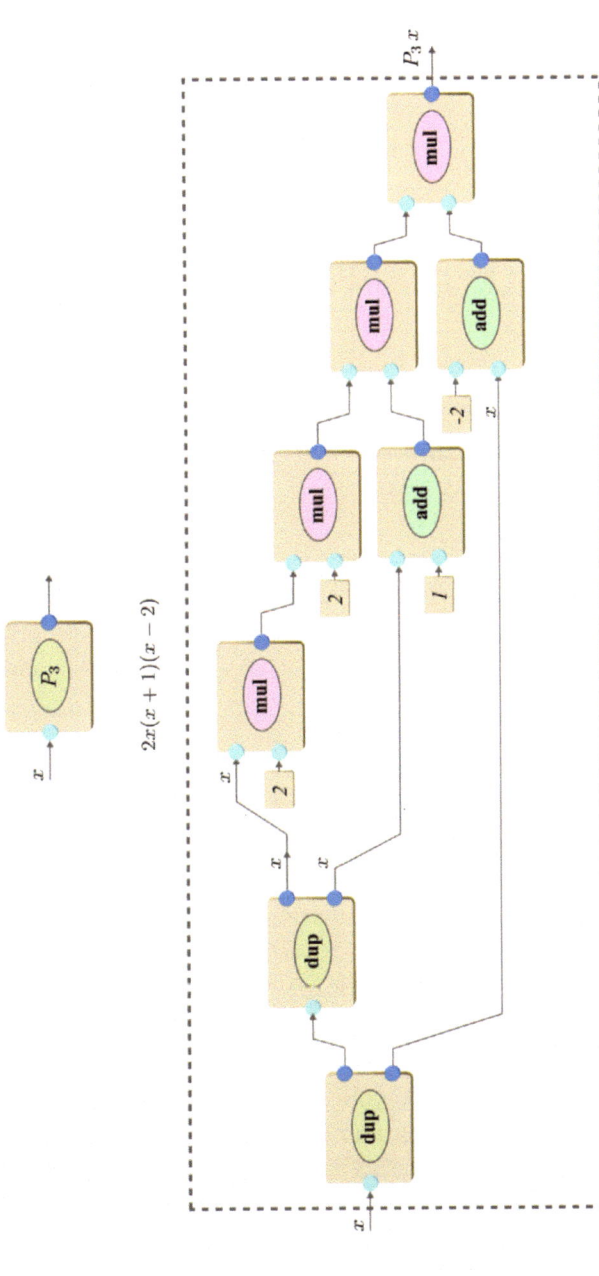

Figure 2.12 Schematic representation of the structure of a cubic polynomial, written in factorized form $2x(x + 1)(x - 2)$.

Figure 2.13 demonstrates a graphical representation of the cubic polynomial $P_3 x$. Obviously, the curve is the same regardless of the way we write the polynomial symbolically. The graphical representation shows that the polynomial has three *roots* – values of the argument x for which the value of the polynomial $P_3 x$ equals zero. It also shows that the polynomial "explodes" at large positive and negative values of the argument.

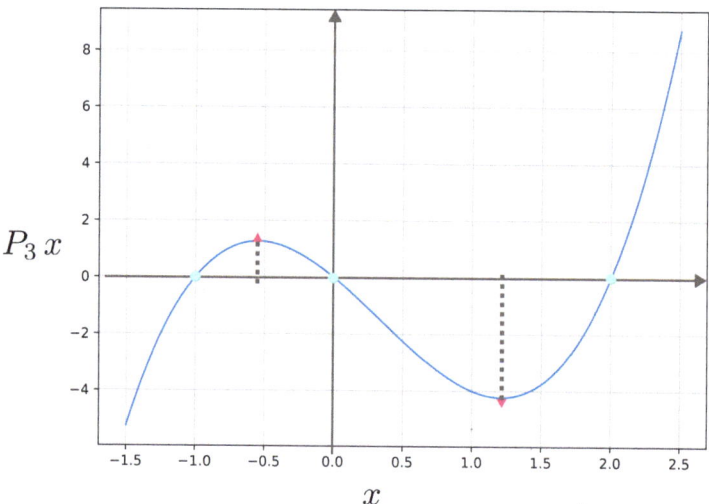

Figure 2.13 Graphical representation of the cubic polynomial $P_3 x = 2x^3 - 2x^2 - 4x$.

Neither graphical nor schematic representation provides an easy way to find the positions of the *local extrema* – points where the polynomial changes its behavior from increasing to decreasing or vice versa. These are marked as upward- or downward-oriented triangles in the Figure 2.13. The symbolic representation, together with the methods of calculus, quickly yields the desired result (the details of how calculus achieves this are of no importance for the point being made.)

We can now appreciate the power of the abstract algebraic ways of representing polynomials. A general polynomial of the degree n can be written as follows:

$$P_n x = a_0 x^0 + a_1 x^1 + a_2 x^2 + \ldots + a_n x^n. \tag{2.5}$$

The reason for writing the first lowest terms

$$a_0 + a_1 x = a_0 x^0 + a_1 x^1$$

will become more clear shortly, when we encounter Einstein's summation rule.

The expression (2.5) can be written more compactly:

$$P_n x = \sum_{i=0}^{i=n} a_i x^i.$$

Here we used a new notational tool – the summation sign $\sum$ (capital Greek letter sigma Σ). The expression above should be read as follows: The function $P_n x$ consists of the sum of terms $a_i x^i$ where the index i takes the values from zero to n, and appears as the subscript of the coefficients a and as the power of the variable x.

The notation using $\sum$ is an improvement compared to the formula (2.5), but there exists an even better way to write various sums of terms with repeated indices. This new and better notation is called *Einstein's summation rule*. It is extremely helpful in vector and tensor mathematics and we consider this notation next.

2.5 EINSTEIN'S SUMMATION RULE

If there existed a Nobel Prize for notation one would definitely go to Albert Einstein for what is known as *Einstein's summation rule*. In short, the rule states: *Whenever you see repeated indices in an expression, understand the expression as the sum of many terms with similar structure.* For example:

$$\boxed{a_i x_i = a_1 x_1 + a_2 x_2 + a_3 x_3 + \ldots}$$

It is a more efficient alternative to the traditional notation:

$$\sum_{i=1}^{i=n} a_i x_i = a_1 x_1 + a_2 x_2 + a_3 x_3 + \ldots + a_n x_n.$$

Already in this simple case, we saved at least 7 symbols: the summation sign $\sum$ and the lower and upper limits $(i = 1)$ and $(i = n)$.

Exercise 2.5 *Using Einstein's summation rule, write the expression for the general form of a polynomial of degree n.*

The last problem reveals that we have to specify the range of values for the summation index. For example:

$$a_i x_i \quad i = 1, 2, \ldots, n.$$

The range for the summation index (or indices if there are several) is usually clear from the context and it is the same across many expressions written using Einstein's summation rule.

Exercise 2.6 *Assuming the summation indices i, j, and k range from 1 to 4, write out fully the following expressions:*

$$(a)\, b_i y_i \qquad (b)\, b_j y_j \qquad (c)\, b_k y_k.$$

The last problem demonstrates an important property of Einstein's summation notation: *We are free to rename the summation indices.* Renaming the indices does not affect the resulting value and can often be useful for manipulating expressions. For example, given the expression

$$a_i x_i + b_j x_j,$$

we can immediately rewrite it as follows:

$$a_i x_i + b_j x_j = a_i x_i + b_i x_i = (a_i + b_i) x_i.$$

Sometimes we need to be careful with new notation:

$$(a_i x_i)^2 \neq a_i^2 x_i^2.$$

Indeed, the left-hand side must be understood as

$$(a_i x_i)(a_i x_i) = (a_i x_i)(a_j x_j) = a_i a_j x_i x_j.$$

Clearly, there will be terms like $a_1 a_2 x_1 x_2$ which are absent from the sum $a_i^2 x_i^2$. The next exercise helps to see this better.

Exercise 2.7 *Write out fully the expressions*

$$(a)\, (a_i x_i)(a_j x_j) \qquad (b)\, a_i^2 x_i^2$$

for $i = 1, 2$.

Exercise 2.8 *Rewrite the expression*

$$(a_i x_i)^2 = \frac{b_j y_j}{c_k c_k}$$

in the traditional form, using the summation symbol Σ. Assume that all indices run from 1 to N.

Einstein's summation rule is used in a great number of problems and we will use it extensively.

Limitation of Einstein's Summation Rule

Einstein's summation rule has its limitations. For example, it is *almost never allowed*[10] to write expressions with more than two repeated indices:

$$a_i b_i c_i \quad \longleftarrow \text{NOT allowed!}$$

Traditional notation allows writing such expressions with ease:

$$\sum_{i=1}^{i=n} a_i b_i c_i = a_1 b_1 c_1 + a_2 b_2 c_2 + \ldots + a_n b_n c_n.$$

2.5.0.1 *Practice With ESR*

It is important to become proficient with Einstein's summation rule because it will be used extensively when dealing with vectors and tensors.

Einstein's summation rule (ESR for short) can be combined with indexed objects introduced earlier (see page 20.) In addition to the indexed objects δ_{ij} and ϵ_{ij}, it will be useful to introduce one more indexed object:

$$\Sigma i = 1 \text{ or } \Sigma_i = 1.$$

Although this indexed object appears trivial, it proves useful. For example, it allows writing simple sums concisely:

$$\Sigma_i a_i = a_1 + a_2 + \ldots + a_n.$$

Exercise 2.9 *Write out in full form and then simplify the following expressions:* $(a)\, \delta_{1i} a_i$ $(b)\, \delta_{3k} a_k$ $i, k = 1, 2, 3, 4,$ $(c)\, \epsilon_{1j} a_j$ $(d)\, \epsilon_{3j} a_j$ $i, j = 1, 2, 3, 4.$

The action of Kronecker's delta on another indexed object results in "renaming" the index of the latter. Indeed, from the sum

$$\delta_{ki} a_i = \delta_{k1} a_1 + \delta_{k2} a_2 + \ldots + \delta_{kn} a_n$$

only the terms with $i = k$ remains:

$$\delta_{ki} a_i = a_k.$$

---

[10] Exceptions are possible, but rare and are irrelevant to our goals.

Exercise 2.10 *Using the last result, show that*

$$a_i + a_j = (1 + \delta_{ji})a_i.$$

Double sums can be handled using Einstein's summation rule as well:

$$\left(\sum_{i=1}^{i=n} a_i\right)\left(\sum_{j=1}^{j=n} b_j\right) = \sum_{i=1}^{i=n}\sum_{j=1}^{j=n} a_i b_j = \Sigma_i \Sigma_j a_i b_j \qquad i, j = 1, \ldots, n.$$

Exercise 2.11 (*a*) *Show that*

$$\delta_{ij} a_i b_j = a_i b_i \quad i, j = 1, \ldots, n.$$

(*b*) *Write out fully*

$$\epsilon_{ij} a_i b_j,$$

assuming $i, j = 1, 2$.

Let us wrap up this chapter with a couple of more advanced problems:

Exercise 2.12 (*a*) *Show that*

$$\delta_{ij} \delta_{jk} = \delta_{ik} \quad i, j, k = 1, \ldots, n.$$

(*b*) *Show that*

$$\epsilon_{ij} \epsilon_{jk} = -\delta_{ik} \quad i, j, k = 1, 2.$$

These relations have simple interpretations in terms of operators and their components, discussed in Chapter 4.

Consider a number

$$x = \delta_{ij} \epsilon_{ij}.$$

Using the fact that we can rename summation indices, we can write

$$x = \delta_{ik} \epsilon_{ik} = \delta_{jk} \epsilon_{jk} = \delta_{ji} \epsilon_{ji}.$$

From the definition of the indexed objects δ and ϵ follows that

$$\delta_{ji} = \delta_{ij} \quad \text{and} \quad \epsilon_{ij} = -\epsilon_{ji}.$$

Thus, we obtain

$$x = \delta_{ji} \epsilon_{ji} = -\delta_{ij} \epsilon_{ij} = -x.$$

Therefore, the value of x must be zero due to the different symmetries of the indexed objects with respect to swapping their indices.

Exercise 2.13 *Show that*

$$\epsilon_{ij} a_i a_j = 0.$$

CHAPTER HIGHLIGHTS

- *The power of mathematical concepts and methods increases with the level of abstraction.*

- *Learning new concepts often involves learning new terminology. The latter can create an artificial mental barrier.*

- *"Usual" numbers form a mathematical structure. The structure is revealed through various relations that exist between numbers.*

- *Relations between numbers are expressed using the concept of functions and operations (e.g., addition). Each operation is characterized by its arity – the number of arguments it accepts as an input.*

- *Functions can be represented schematically as boxes with inputs and outputs.*

- *Functions that act on natural numbers can be written using index notation (e.g., $f\,i = f_i$).*

- *Linear functions represent the simplest but still powerful and useful kinds of functions.*

- *Functions can be composed to create new functions.*

- *A function with several inputs is said to be partially applied when not all its inputs are populated.*

- *The same function can be represented in various ways: Graphical, as a symbolic formula, as a table. The function is not reduced to any of its representations.*

- *The power of abstract mathematical thinking comes, in part, from efficient notation. Einstein's Summation Rule (ESR) is a good example of this.*

CHAPTER 3

Arrows and Vectors

I N THE PREVIOUS CHAPTER WE LEARNED ABOUT numbers and various relations between them. As a particular class of relations, we discussed functions. We introduced *binary* and *unary* functions and different ways functions can be combined (*composed*) to produce new functions. We also learned that functions can be represented in various ways and that none of those different representations defines the concept of function completely. Each representation of a function highlighted some important aspect of it.

Vectors, which will be introduced in this chapter, also allow different representations. We will start with a particular model of vector quantities – *arrows*. It is important to remember that while this model illustrates the concept of vectors, it does not define vectors completely. In other words, arrows are particular examples of vectors, but vectors are more than directed line segments.

To arrive at the definition of vectors we must explore their properties more fully. This will be the goal of the current chapter.

3.1 ARROWS

To arrive at the idea of vectors we will start with simple geometrical objects – arrows in a plane, as illustrated in Figure 3.1.

Symbolically, we will denote vectors by placing an arrow over letters:

$$\vec{a}, \vec{b}, \vec{c}, \ldots, \vec{\alpha}, \vec{\beta}.$$

The length of an arrow $\vec{a}$ is denoted by the same letter without an arrow:

$$\text{length } \vec{a} = a.$$

DOI: 10.1201/9781003620365-3

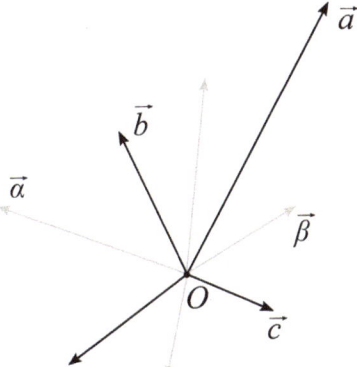

Figure 3.1 Set of arrows starting at the same origin point O. All imaginable arrows taken as one set form the arrow space $\vec{\vec{A}}$.

The set of all possible arrows – called *arrow space* (or *vector space*) – we will denote as

$$\vec{\vec{A}} = \{\vec{a}, \vec{b}, \vec{c}, \ldots, \vec{\alpha}, \vec{\beta} \ldots\}.$$

Diagrammatic representation of the set of arrows is given in the Figure 3.2.

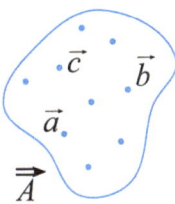

Figure 3.2 All arrows, considered as unified collection of objects of similar nature, can be viewed as a mathematical set. Each arrow is an element of this set, denoted as a point.

Functions on Arrows

The function **length** is a unary function that accepts arrows as input (argument) and returns a numeric value – the lengths of the arrow.
On the level of sets, the **length** function *maps* the set of all vectors into the set of real numbers:

$$\vec{\vec{A}} \overset{\text{length}}{\longrightarrow} \mathbb{R}.$$

In the next chapter, we will encounter other types of functions on arrows. Some of them will return numbers like **length** does, and others will return arrows, mapping *arrow space* into its own copy:

$$\vec{\vec{A}} \longrightarrow \vec{\vec{A}}.$$

Arrows and Numbers

Arrows in a plane include, in some sense, the notion of numbers. Indeed, as shown in Figure 3.3, real numbers may be represented as points on a line – number line. Positive numbers correspond to the arrows pointing to the right of the origin (number zero), and negative numbers correspond to the arrows directed to the left.

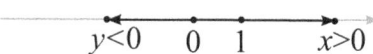

$$y<0 \quad 0 \quad 1 \quad x>0$$

Figure 3.3 Real numbers can be represented by arrows oriented along a fixed line – *number line.*

3.2 ALGEBRA OF ARROWS

Next, we will establish similarities between arrows and numbers by exploring possible *algebraic operations* on vectors.

3.2.1 Combining Arrows

Two arrows can be *combined* in a natural way to yield the third arrow. Using either the head-to-tail approach or a parallelogram method, the vectors $\vec{a}$ and $\vec{b}$ can be combined graphically, as illustrated in Figure 3.4.

Symbolically, we express the combination of two arrows as

$$\kappa\,\vec{a}\,\vec{b} \to \vec{c},$$

where the Greek letter κ (kappa) denotes a binary function that combines arrows. Alternatively, *infix notation* can be used

$$\vec{a}\kappa\vec{b} = \vec{c}.$$

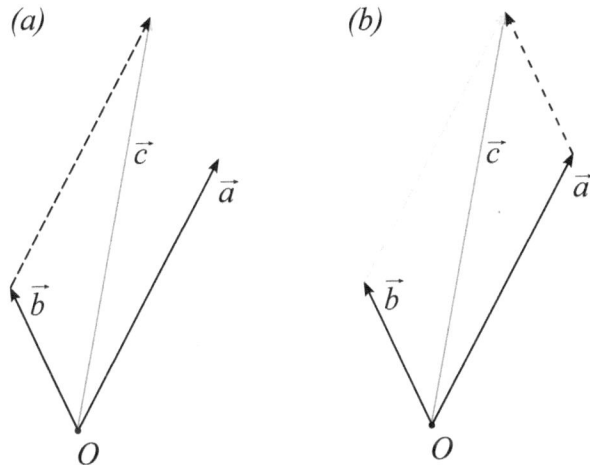

Figure 3.4 Two arrows can be combined to produce a new arrow. One way to do this is to arrange two arrows "tail-to-tip." This operation is called *addition of arrows*: $\vec{a} + \vec{b} = \vec{c}$.

Usually, the same symbol is used for vector "addition" as for addition of numbers:

$$\vec{a} + \vec{b} = \vec{c}.$$

This leads to "double-booking" of the same symbol. In technical jargon, the function "+" is *overloaded* – defined on objects of different kinds (numbers and arrows in this case). We must keep in mind that in the following two expressions:

$$3 + x \qquad \text{and} \qquad \vec{a} + \vec{b}$$

the function "+" has different meanings (it is *implemented* differently as an operation). Ignoring this distinction may lead to syntactically correct but meaningless expressions like this one:

$$3 + \vec{a},$$

which can not be evaluated or computed.

In the spirit of arrow notation, we could write

$$\vec{a} \stackrel{\rightarrow}{+} \vec{b} = \vec{c}.$$

While this notation might be more rigorous and consistent, it is too noisy and laborious to use. We will stick to the traditional overloaded notation, remembering that the meaning is different.

Anti-arrows and special arrow

For every arrow there exists an "anti-arrow" (or *reverse* arrow):

$$\vec{a} \longrightarrow (-\vec{a})$$

with an obvious meaning of $(-\vec{a})$ – arrow of the same length as $\vec{a}$ but pointing in the opposite direction. When added, an arrow and its reverse result in a special element – *zero arrow*:

$$\vec{a} + (-\vec{a}) = \vec{0}.$$

Zero arrow has zero length and may have *any direction* because the latter does not matter in any operations with arrows. Zero arrow plays the same role as the number zero in the algebra of numbers:

$$\vec{b} + \vec{0} = \vec{0} + \vec{b} = \vec{b}$$

for any arrow $\vec{b}$.

So far we established several similarities between numbers and arrows. These similarities concerned addition. How about multiplication?

3.2.2 Arrow Product

Is there an operation on arrows in a plane analogous to the product of two numbers? Let us remind ourselves how a product of two numbers behaves:

$$x * y = y * x$$

and

$$x * (y + z) = x * y + x * z.$$

The product of two numbers is commutative, it is proportional to each of the factors, and the product distributes in a familiar way over the sum.

Now, can we define some operation "$\vec{*}$" or simply "$*$" on any pair of arrows $\vec{a}$ and $\vec{b}$

$$\vec{a} \ \vec{*} \ \vec{b} = \vec{c}$$

such that it produces an arrow *in the same plane as the original vectors* and is proportional to both $\vec{a}$ and $\vec{b}$? Additionally, the product should have a simple distributive property over the sum of two arrows:

$$\vec{a} * (\vec{b} + \vec{c}) = (\vec{a} * \vec{b}) + (\vec{a} * \vec{c}).$$

It is an excellent exercise to come up with various candidates for the product of two arrows and to analyze whether they satisfy this requirement. Here are some example considerations. The variant

$$\vec{a} * \vec{b} = \vec{a}$$

does not work, since the result is not proportional to $\vec{b}$. For similar reasons, the variant $\vec{a} * \vec{b} = \vec{b}$ won't work. The variant

$$\vec{a} * \vec{b} = \vec{a} + 2\vec{b}$$

or any similar sum with arbitrary coefficients won't work, since the results is not proportional to either $\vec{a}$ or $\vec{b}$ (doubling $\vec{a}$ does not double the result; similarly for $\vec{b}$).

A reasonable candidate could be

$$\vec{a} * \vec{b} = \vec{a}b.$$

The result is proportional to a and b, however, this definition won't be compatible with the multiplication properties:

$$\vec{a} * \vec{0} = \vec{a} * (\vec{b} + [-\vec{b}]) = \vec{a} * \vec{b} + \vec{a} * [-\vec{b}] = 2\vec{a}b.$$

This is unacceptable, because we must have

$$\vec{a} * \vec{0} = a * 0 = 0.$$

In later chapters, after we learn about *compound numbers* (Section 6.2) and *linear operators* (Section 4.2), we will encounter a useful definition of a product of two arrows *in a plane* that yields an arrow in *the same plane*.

Arrows and Numbers

Operations with arrows require the concept of numbers. Indeed, without numbers we could not write the following very natural relation:

$$\vec{a} + \vec{a} = 2\vec{a}.$$

The sum of identical arrows yields an arrow twice as long and pointing in the same direction. Numbers are used in many other operations involving arrows and are essential in the algebra of arrows.

3.3 BASIS

We can combine two non-parallel arrows[1]:

$$\vec{a} + \vec{b} = \vec{c}.$$

When read from right to left, this equation states that an arrow $\vec{c}$ can be written as the sum of two other non-parallel arrows. This statement can be generalized: *Any arrow can be "built from" a pair of non-parallel arrows, by "stretching" each arrow if necessary* (see Figure 3.5):

$$\vec{c} = \alpha\vec{a} + \beta\vec{b}.$$

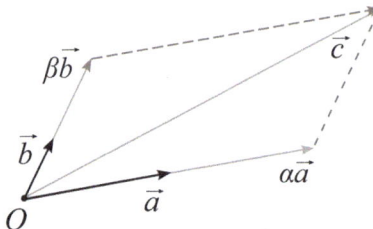

Figure 3.5 A pair of non-parallel arrows, $\vec{a}$ and $\vec{b}$, can be used as basis – building blocks for all other arrows. In this example: $\vec{c} = \alpha\vec{a} + \beta\vec{b}$.

The non-parallel arrows $\vec{a}$ and $\vec{b}$ are called *independent* arrows (one is not the scaled version of the other), and the pair of such arrows is called the *basis*. Once the basis is found and fixed, every arrow gets a numeric *representation* as a pair of real numbers:

$$\vec{c} \longrightarrow (\alpha, \beta).$$

The numbers α and β are called *components* of the arrow $\vec{c}$ *relative to the basis* $\{\vec{a}, \vec{b}\}$.

Note: All we require from the basis arrows is to be independent (non-parallel). They do not have to be perpendicular, and they can have any length. It is often helpful, however, to choose arrows of some unit length and pointing in orthogonal directions. We will return to this point in Sections 3.5 and 5.2.

---

[1]Parallel arrows are combined trivially.

3.3.1 Basis Notation

After some experimentation, it has been established that a special notation for the basis arrows is convenient. All basis arrows are denoted by a letter e. Next, within the given basis the arrows are enumerated using an integer subscript:

$$\vec{e}_1, \vec{e}_2.$$

An arrow $\vec{a}$ is also written in this basis using components with subscripts:

$$\vec{a} = a_1 \vec{e}_1 + a_2 \vec{e}_2.$$

Similarly for any other arrow:

$$\vec{b} = b_1 \vec{e}_1 + b_2 \vec{e}_2.$$

Of course these expressions are for arrows in a plane. For arrows in three dimensions the basis will have three arrows $\{\vec{e}_1, \vec{e}_2, \vec{e}_3\}$ and the expansions will look like

$$\vec{a} = a_1 \vec{e}_1 + a_2 \vec{e}_2 + a_3 \vec{e}_3.$$

This idea can be easily generalized to any number of dimensions!

Exercise 3.1 *Use Einstein's summation rule to write the expansion of an arbitrary vector $\vec{a}$ in terms of the basis vectors $\vec{e}_i, i = 1, \ldots, n$.*

Using components of arrows in a given basis, every graphical operation with arrows can be translated into an algebraic manipulation of components. For example, an addition of three vectors

$$\vec{a} + \vec{b} + \vec{c} = \vec{d}$$

has a component-wise representation (see Fig. 3.6):

$$a_i + b_i + c_i = d_i.$$

This single equation must be understood as "encoding" several equations – one for each value of the index i. Such an approach is very convenient when problems involve "arrows" in three or more dimensions.

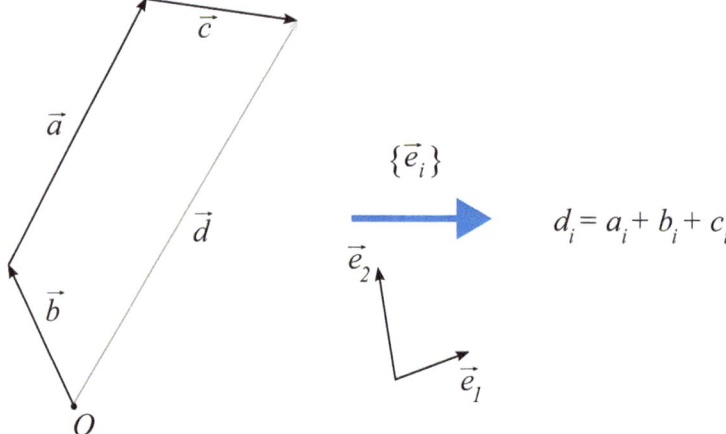

Figure 3.6 Introduction of basis arrows $\{\vec{e}_i\}$ translates operations on arrows from a graphical problem involving manipulation of arrows into an algebraic problem involving manipulation of numbers.

3.4 MANY BASES

Arrows of any basis are neither special nor unique. The choice of basis arrows is arbitrary, with some choices more suitable to a given problem, and some choices less so. Not infrequently several bases are used to analyze the same problem from different "points of view."

To differentiate between various sets of basis arrows they may be labeled with different numbers of "prime" marks:

$$\{\vec{e}_i\}, \quad \{\vec{e}_i'\}, \quad \{\vec{e}_i''\} \quad \text{and so on.}$$

"New" and "Old" Bases

Usually, the basis $\{\vec{e}_i\}$ is introduced first and is called the *old basis* (or *original basis*), whereas the basis $\{\vec{e}_i'\}$ is introduced second and is called the *new basis*.

We can express any arrow in the new basis:

$$\vec{a} = a_1' \vec{e}_1' + a_2' \vec{e}_2' = a_i' \vec{e}_i'.$$

Note that we added the "prime" marks to the components of the arrow a_i' to help identify which basis is used to determine them.

The same arrow is also expressed in terms of the "old"/original basis:

$$\vec{a} = a_i \vec{e}_i.$$

The components a_i and a'_i are related. To find how we can expand the new basis arrows in terms of the old ones:

$$\vec{e}'_1 = E_1 \vec{e}_1 + E_2 \vec{e}_2$$

and

$$\vec{e}'_2 = E_3 \vec{e}_1 + E_4 \vec{e}_2.$$

Here the numbers E_1, E_2, represent the components of $\vec{e}'_1$ relative to the basis $\{\vec{e}_i\}$; similarly for the numbers E_3, E_4.

Rotated Basis

Consider new basis

$$\vec{e}'_1 = \vec{e}_1 + \vec{e}_2 \quad \text{and} \quad \vec{e}'_2 = \vec{e}_1 - \vec{e}_2.$$

In this case $E_1 = E_2 = 1$ and $E_3 = -E_4 = 1$. The basis vector $\vec{e}'_1$ is pointing diagonally between $\vec{e}_1$ and $\vec{e}_2$. Can you find the direction of $\vec{e}'_2$?

More flexible notation for the components of basis arrows is preferred. Instead of using E with a single index, we can use double index notation:

$$\vec{e}'_1 = E_{11} \vec{e}_1 + E_{12} \vec{e}_2$$

and

$$\vec{e}'_2 = E_{21} \vec{e}_1 + E_{22} \vec{e}_2.$$

Note how the first index of E_{ij} matches the basis arrow being expanded, while the second index matches the arrow it multiplies.

After plugging these expressions into the expansion of $\vec{a}$ in terms of $\vec{e}'_1$ and $\vec{e}'_2$, we obtain

$$\vec{a} = a'_1 E_{11} \vec{e}_1 + a'_1 E_{12} \vec{e}_2 + a'_2 E_{21} \vec{e}_1 + a'_2 E_{22} \vec{e}_2.$$

Grouping the terms with respect to the basis arrows, we find

$$\vec{a} = (a'_1 E_{11} + a'_2 E_{21}) \vec{e}_1 + (a'_1 E_{12} + a'_2 E_{22}) \vec{e}_2.$$

Comparing this with the expansion of $\vec{a}$ in terms of the original basis, we obtain:

$$a_1 = a'_1 E_{11} + a'_2 E_{21} = a'_i E_{i1}$$

and

$$a_2 = a'_1 E_{12} + a'_2 E_{22} = a'_i E_{i2}.$$

In a compact form, these two equations become

$$a_j = a'_i E_{ij}.$$

Exercise 3.2 *Using Einstein's summation rule, show that the relations between the "new" and "old" basis arrows can be written as follows:*

$$\vec{e}\,'_i = E_{ij} \vec{e}_j.$$

Power of ESR

The relations between the components of a vector $\vec{a}$ in different bases can be quickly obtained using Einstein's summation rule. Indeed, we have

$$\vec{a} = a_j \vec{e}_j = a'_i \vec{e}\,'_i = a'_i (E_{ik} \vec{e}_k) = a'_i E_{ij} \vec{e}_j,$$

from which follows

$$a_j = a'_i E_{ij}.$$

Contravariant Vectors

An important observation can be made when we compare the relations of components of an arbitrary arrow $\vec{a}$ and the relation between the basis arrows:

$$\vec{e}\,'_i = E_{ij} \vec{e}_j \qquad \text{vs.} \qquad a_j = a'_i E_{ij}.$$

The same set of components E_{ij} allows us to go from the original basis $\vec{e}_j$ to the "new" basis $\vec{e}\,'_i$, and in the *opposite* direction – from a'_i to a_j – for the components of arrows. Due to this "contrary" behavior of the arrow components, arrows are called *contravariant vectors*.

The following exercise helps better understand the contravariant nature of arrows.

Exercise 3.3 *Consider a "new" basis*

$$\vec{e}\,'_1 = \mu\vec{e}_1, \quad \vec{e}\,'_2 = \nu\vec{e}_2.$$

(a) *Write out the components E_{ij} explicitly.*

(b) *Given the coefficients a_1 and a_2 of a vector $\vec{a}$, find the coefficients a'_1 and a'_2 relative to the "new" basis.*

The last problem demonstrates that when basis arrows are scaled ("stretched" or "compressed"), the corresponding coefficients are also scaled, but in the *opposite direction*. It has to be this way if we want the combination $\vec{a} = a_1\vec{e}_1 + a_2\vec{e}_2$ to remain the same and represent *the same arrow* in *different bases*. Thus, the change (variation) of components of arrows is in a certain sense opposite (contrary) to the behavior of the basis arrows, justifying the name *contravariant*.

Covariant and Contravariant Vectors

There are only two kinds of vectors[2]: *contravariant* and *covariant*. As we found out, the components of contravariant vectors are related via the transformation rule

$$a_j = a'_i E_{ij} .$$

Later, in Section 5.6, we will discover new type of vectors (*covariant vectors*) whose components transform similarly to basis vectors (*covariantly*):

$$b'_i = E_{ij} b_j .$$

Here E_{ij} are the components of the "new" basis arrows in terms of the "old" basis arrows:

$$\vec{e}\,'_i = E_{ij}\vec{e}_j .$$

Arrows represent contravariant vectors. We will soon learn what kind of objects correspond to covariant vectors.

3.4.1 Transposition*

At this point a reminder must be given: All material in sections with an asterisk (like this one) is either optional or slightly more advanced than usual. These sections can be skipped without a significant impact on the main message of the book.

---

[2]Although we have not formally defined vectors yet, we can still acknowledge that they are not all the same.

The way of writing the relation between the components of the same vector $\vec{a}$ in different bases

$$a_j = a'_i E_{ij}$$

seems different from the relation between the basis vectors:

$$\vec{e}\,'_i = E_{ij}\vec{e}_j.$$

The difference in notation – the use of j for the components of $\vec{a}$ instead of i – is superficial since the labeling of indices in these expressions is not fixed. Indeed, we can rewrite the former expression as

$$a_j = a'_i E_{ij} \quad \rightarrow \quad a_j = a'_k E_{kj} \quad \rightarrow \quad a_i = a'_k E_{ki}$$

and finally

$$a_i = a'_j E_{ji},$$

keeping the summation of the "primed" component with the first index of E_{ji}.

We can modify the last expression further, writing

$$a_i = F_{ij}a'_j.$$

Here the summation of the components a'_j happens with the *second* index of some coefficients F_{ij}, in full analogy with the relation between the basis vectors. From the equality

$$a'_j E_{ji} = F_{ij}a'_j$$

follows the relations between the coefficients E_{ij} and F_{ij}:

$$F_{ij} = E_{ji}.$$

For a two dimensional plane, these can be written fully:

$$F_{11} = E_{11},\ F_{12} = E_{21},\ F_{21} = E_{12},\ F_{22} = E_{22}.$$

The coefficients F are obtained from E by *transposing* the latter; symbolically we can write

$$F = T(E),$$

or, using a more traditional notation:

$$F = E^T \text{ and } E^T_{ij} = E_{ji}.$$

Matrices

Often the coefficients and components that require two indices, like E_{ij}, are conveniently written using special tables called *matrices*. For example, we can write

$$E = \begin{pmatrix} E_{11} & E_{12} \\ E_{21} & E_{22} \end{pmatrix}.$$

Using the matrix form, the operation of transposition looks like "swapping" of the non-diagonal elements:

$$E^T = \begin{pmatrix} E_{11} & E_{21} \\ E_{12} & E_{22} \end{pmatrix}.$$

Although matrices are often used in practical calculations, they will not be used much in this book.

Using transposition, we can write the relation between the components of any vector in different bases as

$$a_i = E_{ij}^T a_j' \quad \left(\vec{e}_i' = E_{ij} \vec{e}_j \right).$$

Comparing the relations between the components of the vector and between the basis vectors, we see that the indices are labeled identically, and the summation happens in a similar manner – using the second index of the coefficients.

This form of writing the relations between the components of vectors might appear more consistent or aesthetically pleasing, but it has no significant mathematical advantage. We did this exercise with the sole purpose of introducing an important *operation of transposition*. Transposition is used often in the world of vectors and tensors.

3.4.2 Going The Opposite Way

Above, we expanded the "new" basis in terms of the "old" one: $\vec{e}_i' = E_{ij} \vec{e}_j$. We can expand the "old" basis $\vec{e}_i$ in terms of the "new":

$$\vec{e}_i = E_{ij}' \vec{e}_j'.$$

The components E_{ij}' are related to the components E_{ij}. We will see that they "cancel" each other in a certain sense. Indeed, if we plug into the last equation the expansion

$$\vec{e}_j' = E_{jk} \vec{e}_k,$$

we will arrive at the expression

$$\vec{e}_i = E'_{ij} E_{jk} \vec{e}_k.$$

Exercise 3.4 *Write out explicitly the sum implied in $E'_{ij} E_{jk}$, assuming $i = 1$ and $k = 2$.*

Given the obvious fact that

$$\vec{e}_1 = 1\,\vec{e}_1 + 0\,\vec{e}_2,$$

$$\vec{e}_2 = 0\,\vec{e}_1 + 1\,\vec{e}_2,$$

we conclude that the components E_{jk} and E'_{ij} satisfy the relation

$$\boxed{E'_{ij} E_{jk} = \delta_{ik},} \tag{3.1}$$

where δ_{ik} is the already familiar Kronecker's delta.

The relations (3.1) represent 4 equations with 4 unknowns. This means that if we know the components E_{ij}, we can calculate the components E'_{mn} (unknowns) and vice versa. Indeed, the relations (3.1) will be explicitly written out as follows:

$$
\begin{aligned}
E'_{1j} E_{j1} &= 1, \\
E'_{1j} E_{j2} &= 0, \\
E'_{2j} E_{j1} &= 0, \\
E'_{2j} E_{j2} &= 1.
\end{aligned}
$$

If we denote (taking E_{ij} as known in this case)

$$a = E_{11}, b = E_{12}, c = E_{21}, d = E_{22},$$

and (taking E'_{ij} as unknown in this case)

$$w = E'_{11}, x = E'_{12}, y = E'_{21}, z = E'_{22},$$

then the set of four equations becomes

$$
\begin{aligned}
aw + cx &= 1, & (3.2) \\
bw + dx &= 0, & (3.3) \\
ay + cz &= 0, & (3.4) \\
by + dz &= 1. & (3.5)
\end{aligned}
$$

Exercise 3.5 *Solve the equations (3.2-3.5) and show that*

$$w = d/\Delta, \ x = -b/\Delta, \ y = -c/\Delta, \ z = a/\Delta,$$

where $\Delta = ad - bc$.

Hint: *Notice how the four equations can be grouped into two independent sets of two equations with two unknowns.*

We showed that the components E_{ij} and E'_{mn} satisfy the relations

$$E'_{ij} E_{jk} = \delta_{ik}.$$

There are four more useful relations that the components E_{ij} and E'_{mn} satisfy. To find them, we must expand the "new" basis arrows in terms of the "old" ones:

$$\vec{e}\,'_i = E_{ij} \vec{e}_j,$$

and then expand the "old" arrows in terms of the "new" basis, to obtain

$$\vec{e}\,'_i = E_{ij} E'_{jk} \vec{e}\,'_k.$$

Using the same arguments as before, we conclude that

$$\boxed{E_{ij} E'_{jk} = \delta_{ik}.} \tag{3.6}$$

These relations are not the same as (3.1), since the order of E and E' is switched. If we try to switch the order of E and E' in (3.6)

$$E'_{jk} E_{ij} = \delta_{ik}$$

we will discover that the summation indices differ from the summation indices in (3.1). Thus, there are two ways to express the fact that the components E and E' "cancel" each other: $EE' = 1$ and $E'E = 1$.

The relationship between the coefficients E_{ij} and E'_{ij}, as well as the difference between covariant and contravariant vectors is summarized in Figure 3.7.

Exercise 3.6 *Starting with the relations (3.6) and using the same notation for the known and unknown components, write down the four equation with four unknowns w, x, y, z, and then solve them in terms of the known components a, b, c, d. Prove that the values of the coefficients E'_{ij}, determined by the relations (3.6) are the same as those determined by the relations (3.1).*

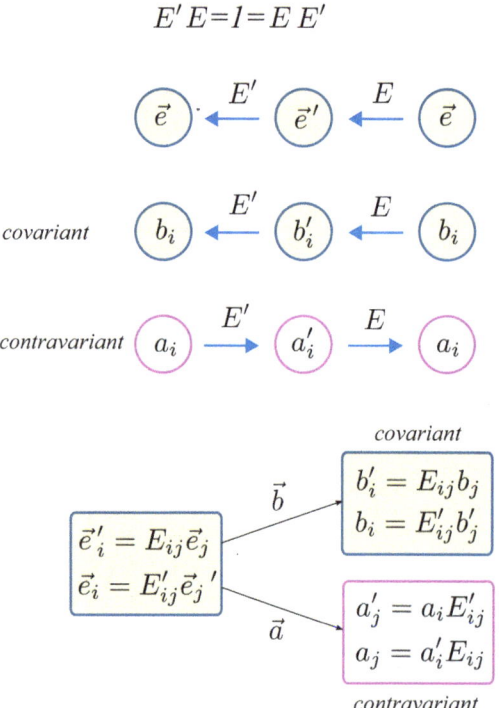

Figure 3.7 Summary of relations between components for covariant and contravariant vectors. Components of covariant vectors change in the same way as the basis vectors. Components of contravariant vectors change in the way opposite to the change of the basis vectors.

3.5 ORTHONORMAL BASES

When basis arrows have unit length and point in mutually perpendicular directions, the basis is called *normal* (normalized length) and *orthogonal*, or just *orthonormal* basis. There are infinitely many such bases and one example is shown in Figure 3.8. Orthonormal bases are very convenient because many calculations involving components are performed more efficiently and many expressions look simpler in such bases (as an example see *scalar product* in the Section 5.2).

3.6 VECTORS DEFINED

Now we are ready to define the concept of a vector, or rather vectors since a single vector is both useless and hard to define.

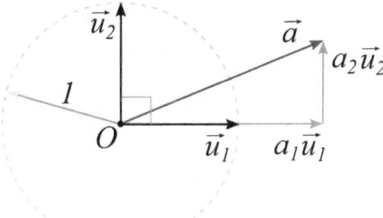

Figure 3.8 Orthogonal arrows of unit length form *orthonormal basis.* Here we denoted the basis arrows using the letter u instead of e to emphasize their unit length.

Vectors

Vectors are *mathematical objects* with the following essential properties:

- Vectors can be combined (added) pairwise to yield another vector.

- Vectors can be multiplied by real numbers to yield vectors of parallel direction but of different length (magnitude or scale).

- There must be at least one basis, and all vectors can be represented via components in the specified basis.

- When the basis changes, components transform *in a very specific way*, to ensure that the *vector remains the same*.

There are two kinds of vectors: *covariant* and *contravariant*. Arrows and arrow-like quantities correspond to contravariant vectors.

We will see that arrows and arrow-like quantities are not the only entities that behave like vectors. In the next chapter, we will learn about various functions on vectors (called *operators*) and understand how some of them represent *covariant vectors*.

CHAPTER HIGHLIGHTS

- *Arrows in a plane provide a simple model for vectors.*

- *Arrows can be manipulated in ways analogous to numbers: Two arrows can be added, and an arrow can be "scaled" (stretched or compressed). Arrows form an algebra.*

- *Basis is an extremely important concept. Basis is a set of objects (arrows) that can be used to "build" all other similar objects (arrows). At the same time, basis can not be used to build itself – basis arrows are independent.*

- *Basis can be chosen in an infinite number of ways. There is no special basis. Different bases might be useful for different problems.*

- *Given a basis, arrows can be specified by writing their components (using index notation) relative to the basis.*

- *Einstein's summation rule is very useful for manipulating expressions involving components of arrows.*

- *The exact values of the arrow's components depend on the basis. Changing the basis changes the values of components, while the arrow remains the same. This is one of the defining properties of vectors.*

- *When the basis is changed, components of the same arrow transform in a very specific way. Depending on the exact form of transformation, we can speak of two kinds of vectors: contravariant and covariant. Arrow-like vectors are examples of contravariant vectors.*

CHAPTER 4

Operators

O OPERATOR CONCEPT EXTENDS THE IDEA OF A
function. An unary numeric function f takes some numeric value
x as an input and produces another numeric value y:

$$f\,x = y \quad \text{or} \quad x \xrightarrow{\;f\;} y.$$

In mathematical jargon, f *maps* x into y.

Similarly, we can study functions that *operate* on arrows/vectors,
yielding other arrows/vectors. The parallel between unary functions over
numbers and unary operators over vectors is highlighted in Figure 4.1.

4.1 OPERATORS ON ARROWS

An action of an operator F on arrows can be represented symbolically
as an equation:

$$F\,\vec{a} = \vec{b}\,.$$

Often a "hat" is placed on top of an operator[1], to emphasize that it is
different from a numeric function:

$$\boxed{\widehat{F}\,\vec{a} = \vec{b}\,.}$$

Notice that the argument of an operators *is not* placed inside parentheses. This is a standard convention.

---

[1]In Quantum Physics, for example.

DOI: 10.1201/9781003620365-4

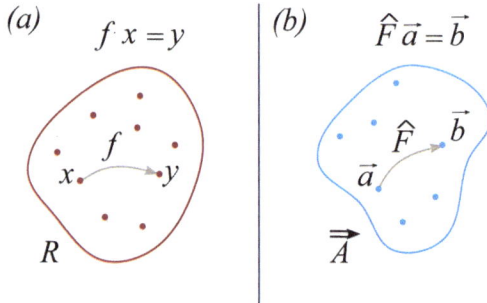

Figure 4.1 Operators extend the idea of functions. (a) An unary function f can be applied to a number x to produce another number y. (b) An unary operator $\widehat{F}$ can be applied to a vector $\vec{a}$ to yield another vector $\vec{b}$.

Simple Operators

It is easy to come up with examples of operators:

- Unit operator (or *identity* operator) $\widehat{I}$, such that

$$\widehat{I}\vec{a} = \vec{a}.$$

- "Zeroing" operator $\widehat{0}$ that maps every vector into a zero vector:

$$\widehat{0}\vec{a} = \vec{0}.$$

- Scaling operator $\widehat{S}_5$, such that
$$\widehat{S}_5\,\vec{a} = 5\vec{a}.$$

- Rotation operator $\widehat{L}$, such that
$$\widehat{L}\vec{a} = \vec{b},$$

where $\vec{b}$ is rotated by $45°$ counter-clockwise relative to the vector $\vec{a}$.

To fully describe an operator $\widehat{F}$ we must describe how it acts *on every* arrow. In general, this requires an *infinite* amount of information, since the number of arrows is infinite and the action of $\widehat{F}$ on different arrows might be different[2]. Describing the action of a general operator in graphical terms using arrows is a hopeless task. In this situation the component representation of arrows saves the day. Let's see how it works.

If we fix a basis $\{\vec{e}_i\}$, then every vector gets an algebraic representation as a set of components:

$$\vec{a} \quad \xrightarrow{\{\vec{e}_k\}} \quad a_i,$$

$$\vec{b} \quad \xrightarrow{\{\vec{e}_k\}} \quad b_j.$$

Now to describe the action of an operator $\widehat{F}$ on the vector $\vec{a}$ we can specify the components of the result b_j for arbitrary components of the input a_i. For vectors in a plane, we have

$$b_1 = F_1 \, a_1 \, a_2$$

and

$$b_2 = F_2 \, a_1 \, a_2.$$

Thus, a pair of binary numeric functions F_1 and F_2 is sufficient to describe an operator. A situation is significantly simplified in the case of a very important class of *linear operators* which we will discuss soon.

Examples

Let us take a closer look at a couple of operators. While studying these examples we must keep in mind that the relations between components are *specific to basis* and will change if we change the basis. The question of how exactly the relation between components changes will be addressed later in Section 4.7 for the simplest types of operators.

The first example operator $\widehat{F}$ acts on the components as follows:

$$b_1 = a_1 + a_2,$$

$$b_2 = a_1 * a_2.$$

To illustrate these relations visually, we can start with arrows of equal lengths but pointing in all directions. The tips of such arrows will lie on a circle, as shown in Figure 4.2 using blue dotted lines. The tips of

---

[2]Simple operators given before are special cases when it is easy to describe the action of operators on *all* vectors.

$$b_1 = a_1 + a_2, \quad b_2 = a_1 * a_2$$

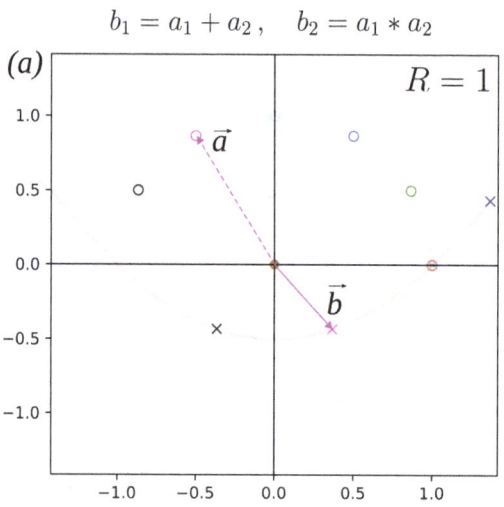

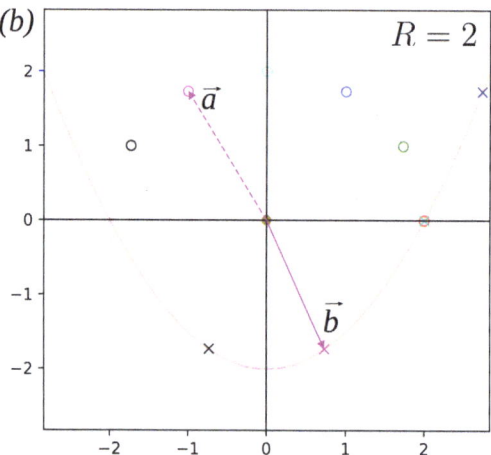

Figure 4.2 The action of the operator $\widehat{F}$ on planar arrow-vectors $\vec{a}$. To demonstrate how $\widehat{F}$ acts on arrows of different directions and lengths, we consider what happens to the circles of two different radii.

all output arrows $\vec{b} = \widehat{F}\vec{a}$ lie on a curve shown in Figure 4.2 using a solid red line. In the first example, the arrows tracing the circle of the radius $R = 1$ are transformed into arrows tracing a curve that looks like a parabola.

Exercise 4.1 *Show that when the components of the output arrow $\vec{b}$ are given by*

$$b_1 = a_1 + a_2,$$

$$b_2 = a_1 * a_2,$$

then the circle with the radius R becomes a parabola described by the equation

$$b_2 = b_1^2/2 - R^2/2.$$

In the second example, the operator $\widehat{G}$ acts on the components as follows:

$$b_1 = a_1 - a_2,$$

$$b_2 = -a_1^2 * a_2.$$

The effect of this operator on the arrows forming a circle is illustrated in Figure 4.3. It seems that increasing the radius of the circles does not substantially change its "image" (solid red line) and only "stretches" the output curve both horizontally and vertically.

Linear and Nonlinear Operators

The operators $\widehat{F}$ and $\widehat{G}$ are examples of rather complicated operators. They are *nonlinear* because they lack the simple property of *linearity*, defined previously in subsection 2.3.3.

Linear operators connect components of input and output vectors in a simple way:

$$b_1 = A\,a_1 + B\,a_2, \quad b_2 = C\,a_1 + D\,a_2.$$

Here $A, B, C,$ and D are numbers. Different linear operators differ only in the values of these numbers. Despite their simplicity, linear operators are powerful and widely used.

Since two arrows can differ only in their magnitude (length) and directions, the action of an operator can be represented by a combination of two steps: rotation and scaling; see Figure 4.4. The order of scaling and rotation does not matter:

$$\widehat{F} = \widehat{S} \circ \widehat{R} = \widehat{R} \circ \widehat{S}.$$

As our next step, we will study rotation operators in more details.

$$b_1 = a_1 - a_2, \quad b_2 = -a_1^2 * a_2$$

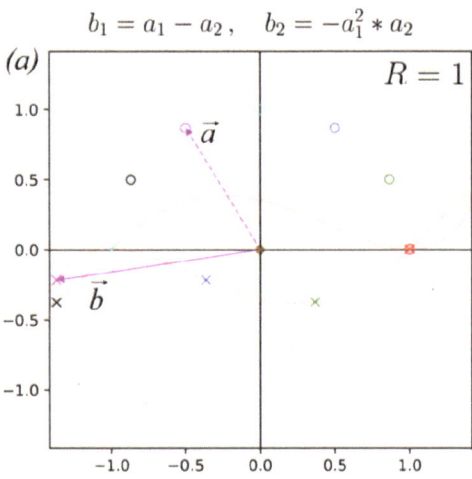

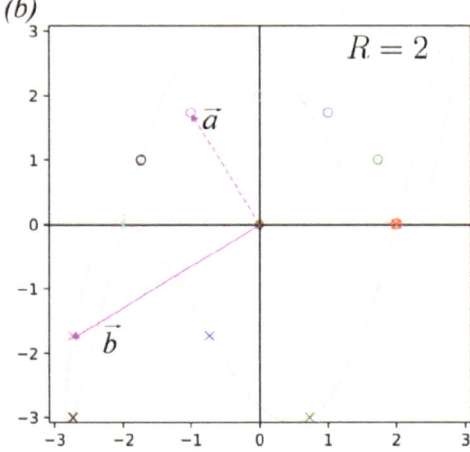

Figure 4.3 An action of an operator $\widehat{G}$ on a planar vector $\vec{a}$. To demonstrate how $\widehat{G}$ acts on arrows of different directions and lengths, we consider what happens to the circles of two different radii.

4.1.1 Rotation Operator

A simple non-trivial operator[3] rotates a vector by some angle, as demonstrated in the Figure 4.5. Strictly speaking, there are infinite number of such operators, one for each rotational angle θ.

---

[3] An example of a trivial operator is the identity operator $\widehat{I}$ which does not change the input vector: $\widehat{I}\vec{a} = \vec{a}$. Another example is the operator that always returns "zero"-arrow: $\widehat{0}\vec{a} = \vec{0}$.

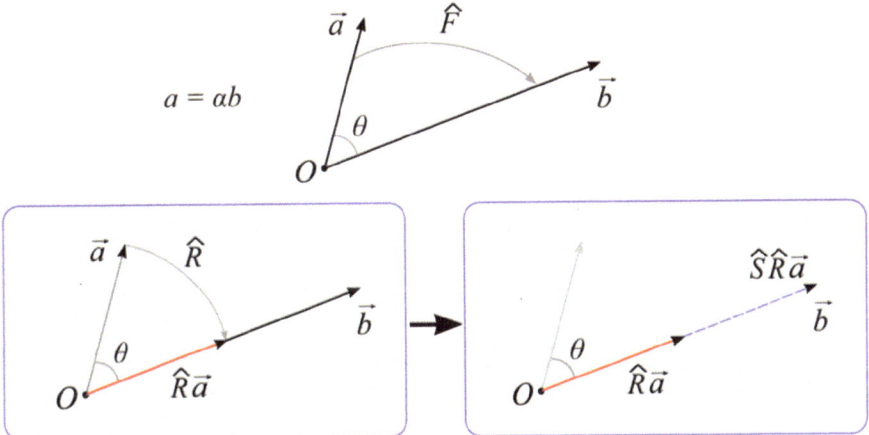

Figure 4.4 An action of an operator $\widehat{F}$ on a planar vector $\vec{a}$ can be described by the sequence of rotation $\widehat{R}$ and scaling $\widehat{S}$: $\widehat{F} = \widehat{S} \circ \widehat{R}$.

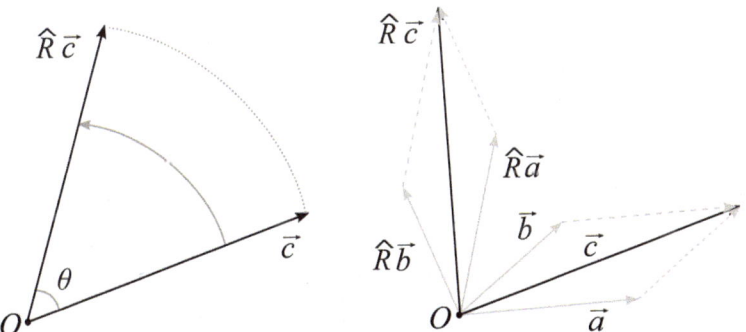

Figure 4.5 The action of the rotation operator $\widehat{R}$ on a planar vector $\vec{a}$ results in the rotation by a specified angle θ. Such rotation possesses an important property: $\widehat{R}(\vec{a} + \vec{b}) = \widehat{R}\vec{a} + \widehat{R}\vec{b}$.

An important property of any rotation operator is that it preserves some relations between arrows. For example, given a vector

$$\vec{c} = \vec{a} + \vec{b},$$

its "image" $(\widehat{R}\,\vec{c})$ can be constructed from rotated vectors $(\widehat{R}\,\vec{a})$ and $(\widehat{R}\,\vec{b})$:

$$\widehat{R}\,\vec{c} = \widehat{R}\,(\vec{a} + \vec{b}) = (\widehat{R}\,\vec{a}) + (\widehat{R}\,\vec{b}).$$

This property is the consequence of two features of the rotation operator: 1) it does not change the length of a vector; 2) it rotates every vector by the same amount, thus keeping the relative angle between any two vectors intact, as illustrated in Figure 4.5.

Another important property of rotation operators is an obvious one:

$$\widehat{R}(\alpha\vec{a}) = \alpha(\widehat{R}\,\vec{a}).$$

Rotation operators represent a subset of a larger set of important operators – *linear operators*.

4.2 LINEAR OPERATORS

Linear operators are operators with a simple yet important property of *linearity*. Two things are required for a linear operator $\widehat{L}$:

$$\boxed{\widehat{L}\,(\vec{a} + \vec{b}) = (\widehat{L}\,\vec{a}) + (\widehat{L}\,\vec{b})}$$

and

$$\boxed{\widehat{L}\,(\alpha\vec{a}) = \alpha(\widehat{L}\,\vec{a}).}$$

Rotation operators satisfy these requirements, as we established in the previous section.

To represent linear operators numerically using basis arrows requires relatively little information. Indeed, for a general arrow $\vec{a} = a_1\vec{e}_1 + a_2\vec{e}_2$ the action of the operator $\widehat{L}$ can be written as

$$\widehat{L}\,(a_1\vec{e}_1 + a_2\vec{e}_2) = [\widehat{L}\,(a_1\vec{e}_1)] + [\widehat{L}\,(a_2\vec{e}_2)] = a_1(\widehat{L}\,\vec{e}_1) + a_2(\widehat{L}\,\vec{e}_2).$$

Thus, to define the action of the linear operator $\widehat{L}$ on an arbitrary arrow $\vec{a}$, it is sufficient to define its action on the basis arrows $\vec{e}_1$ and $\vec{e}_2$.

Operator $\widehat{L}$ acts on arrows to yield other arrows. Thus, we can write

$$\widehat{L}\,\vec{e}_1 = \vec{f}_1 \text{ and } \widehat{L}\,\vec{e}_2 = \vec{f}_2.$$

Like any other vectors, the vectors $\vec{f}_1$ and $\vec{f}_2$ can be written in terms of the basis vectors:

$$\vec{f}_1 = l_1\vec{e}_1 + l_2\vec{e}_2,$$
$$\vec{f}_2 = l_3\vec{e}_1 + l_4\vec{e}_2.$$

These equations can be written more compactly if we improve the notation. Firstly, instead of numbers l_1, l_2, l_3, l_4 we will write

$$\widehat{L}\,\vec{e}_1 = L_{11}\vec{e}_1 + L_{12}\vec{e}_2$$

and

$$\widehat{L}\,\vec{e}_2 = L_{21}\vec{e}_1 + L_{22}\vec{e}_2.$$

Notice how the subscript indices match nicely with the indices of the basis arrows: The first subscript index of L_{ij} corresponds to the basis vector being acted on, while the second index corresponds to the basis vectors being multiplied by L_{ij}. The second step is to use the summation agreement:

$$\widehat{L}\,\vec{e}_1 = L_{1j}\vec{e}_j,$$
$$\widehat{L}\,\vec{e}_2 = L_{2j}\vec{e}_j.$$

Finally, we can write the most compact form:

$$\boxed{\widehat{L}\,\vec{e}_i = L_{ij}\vec{e}_j.} \qquad (4.1)$$

In summary, the action of a linear operator $\widehat{L}$ on the basis arrows (and, consequently, on *any* arrow) is completely determined by its components L_{ij} — just four numbers for arrows in a plane[4].

The components of a linear operator $\widehat{L}$ are specific to the basis. This is completely analogous to the components of arrow-vectors. Indeed, the use of a basis translates arrows and linear operators from the graphical world of drawings into the algebraic world of numbers:

$$\vec{a} \xrightarrow{\{\vec{e}_1,\vec{e}_2\}} a_i,$$
$$\widehat{L} \xrightarrow{\{\vec{e}_1,\vec{e}_2\}} L_{ij}.$$

---

[4]In general, for a space of N dimensions the number of components equals N^2.

Simple Linear Operators

Four simple linear operators can be defined without specifying their components:

- Unit operator, such that $\widetilde{I}\vec{a} = \vec{a}$.

- "Zeroing" operator that maps every vector into a zero vector:

$$\widehat{0}\vec{a} = \vec{0}.$$

- Scaling operators, such that $\widehat{S}\vec{a} = \alpha\vec{a}$ for some specified value α.

- Rotation operators, such that $\widehat{R}_\theta\,\vec{a} = \vec{b}$, where $\vec{b}$ is simply rotated by specified angle θ.

To find the components of any operator we must see how it acts on basis vectors. Let's see how this works for the scaling operator $\widehat{S}$ described above:

$$\widehat{S}\vec{e}_1 - \alpha\vec{e}_1 + 0\vec{e}_2,$$

$$\widehat{S}\vec{e}_2 = 0\vec{e}_1 + \alpha\vec{e}_2.$$

From these equations, we can read the components S_{ij} and write them in matrix form:

$$\widehat{S} = \begin{pmatrix} S_{11} & S_{12} \\ S_{21} & S_{22} \end{pmatrix} = \begin{pmatrix} \alpha & 0 \\ 0 & \alpha \end{pmatrix}.$$

A similar approach can be used to find the components of any linear operator.

Exercise 4.2 *Consider an operator $\widehat{N}$ which "normalizes" an arrow – returns an arrow of unit length pointing in the same direction as the original one. For example:*

$$\widehat{N}\vec{a} = \vec{u}_a = \frac{\vec{a}}{a}.$$

Is it a linear operator?

$\widehat{J}$-operator

Operator that rotates any vector in a plane by 90 degrees counter-clockwise has a special notation (it will be heavily used in Sections 6.2 and 6.3):

$$\widehat{R}_{\pi/2} = \widehat{J}.$$

It is instructive to see how this operator acts on an orthonormal basis $\{\vec{e}_i\}$:

$$\widehat{J}\vec{e}_1 = \vec{e}_2 = 0\,\vec{e}_1 + 1\,\vec{e}_2$$

and

$$\widehat{J}\vec{e}_2 = -\vec{e}_1 = -1\,\vec{e}_1 + 0\,\vec{e}_2.$$

From the last two expressions we can find the components of the $\widehat{J}$-operator:

$$J_{ij} = \begin{pmatrix} J_{11} & J_{12} \\ J_{21} & J_{22} \end{pmatrix} = \begin{pmatrix} 0 & 1 \\ -1 & 0 \end{pmatrix}.$$

Here we used *matrix* form of presenting the components of linear operators. In this form, the first index of the components corresponds to the row in the matrix-table, and the second index corresponds to the column.
Another important property of this operator can be seen when we act on any vector twice:

$$\widehat{J}(\widehat{J}\vec{a}) = -\vec{a}.$$

Indeed, rotating any vector twice by 90 degrees results in total rotation by 180 degrees – the direction of the original vector is flipped. What we showed is that the sequence $\widehat{J} \circ \widehat{J}$ is the same as the operator $(-\widehat{I})$:

$$\boxed{\widehat{J} \circ \widehat{J} = -\widehat{I}.}$$

If you are familiar with complex numbers and know that $i^2 = i * i = -1$, you might see the analogy between the $\widehat{J}$-operator and the "imaginary unit" i. The analogy is not accidental and will be explored in detail in Section 6.2.

4.3 PLOTTING LINEAR OPERATORS

A numeric unary function $f\,x = y$ can be represented graphically as a plot "$y\,vs\,x$." In this plot, we indicate a value y *for each value of* x.

Linear operators of the type

$$\widehat{L}\vec{a} = \vec{b}$$

also allow graphical representation. We can, for example, draw an arrow $\vec{b}$ *for each value of* $\vec{a}$. The input vector $\vec{a}$ can vary both its lengths and direction. For a linear operator $\widehat{L}$ the change in length of the input vector $\vec{a}$ is handled trivially:

$$\widehat{L}(\alpha\vec{a}) = \alpha(\widehat{L}\vec{a}).$$

In words: To find the action of a linear operator on a scaled vector we first apply the operator to the original vector and then scale the result. It also follows that all vectors $\widehat{L}(\alpha\vec{a})$ corresponding to different values of the scale factor α are *parallel* – they are all parallel to the vector $\vec{b} = \widehat{L}\vec{a}$.

The preceding considerations show that to describe the action of a linear operator on various input vectors we can focus only on vectors with unit lengths, but pointing in all possible directions. The tips of all such vectors form a unit circle, as illustrated in Figure 4.6.

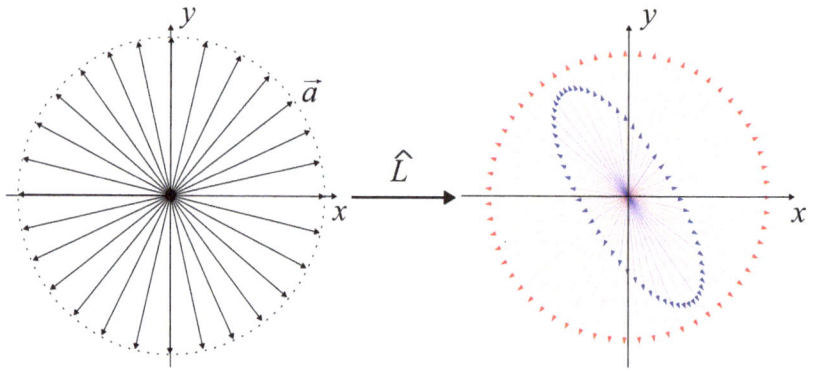

Figure 4.6 Input vectors for a linear operator can vary in direction. In this example, the components of the operator are: $L_{11} = 2/10, L_{12} = 3/10, L_{21} = 5/10, L_{22} = -7/10$.

For linear operators, it is sufficient to plot only the half of the circle, since

$$\widehat{L}(-\vec{a}) = -(\widehat{L}\vec{a}).$$

In other words, the missing half is the inverted copy of the original half. An illustration of this is given in Figure 4.7 for an operator $\widehat{L}$ with components $L_{11} = 2/10, L_{12} = 3/10, L_{21} = 5/10, L_{22} = -7/10$.

Another linear operator, with components $L_{11} = 5/10, L_{12} = 2, L_{21} = 0, L_{22} = 3/2$, is graphically represented in Figure 4.8. In the Figure 4.8(a)

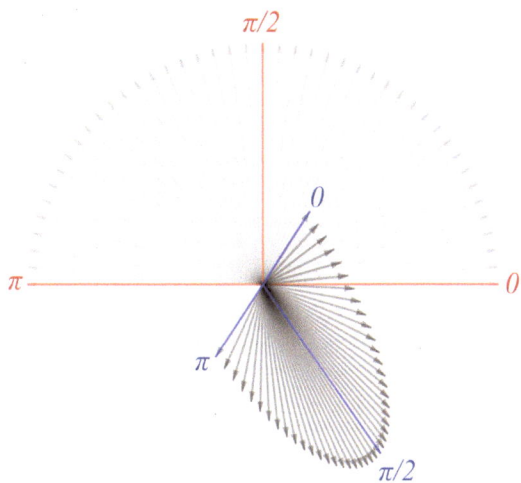

Figure 4.7 To graphically describe a linear operator we can specify how it acts on vectors from the top half of the circle (the bottom is found by inverting the top). In this example, the components of the operator are: $L_{11} = 2/10, L_{12} = 3/10, L_{21} = 5/10, L_{22} = -7/10$.

the action of the operator only on unit vectors from the top half-plane is shown, with the missing part (bottom half-plane) is easily constructed by inverting the transformed part through the origin. Although not easily seen, there are two pairs of vectors[5] that are transformed rather simply by this linear operator – they are scaled without rotation. Two such vectors are shown in Figure 4.8(b). Such vectors are called *eigen-vectors* of a given operator. Eigen-vectors are discussed in more detail in the next section.

Another way to graphically represent a linear operator is to plot two functions: 1) How much an input unit vectors gets rotated; 2) How much an input unit vector gets stretched. This is done in Figure 4.8(c) and (d). From the Figure 4.8(c) it can be seen that there are two unit vectors that are not rotated by the operator (rotation angle is zero for the input vectors at 90 degrees and at about 130 degrees or 310=130+180 degrees).

Eigen-vectors are important and finding them for a given linear operator is an problem often encountered in physics and mathematics. This problem is called *eigen-vector problem* or *eigen-problem* for short.

---

[5]Two vectors and their reversed images.

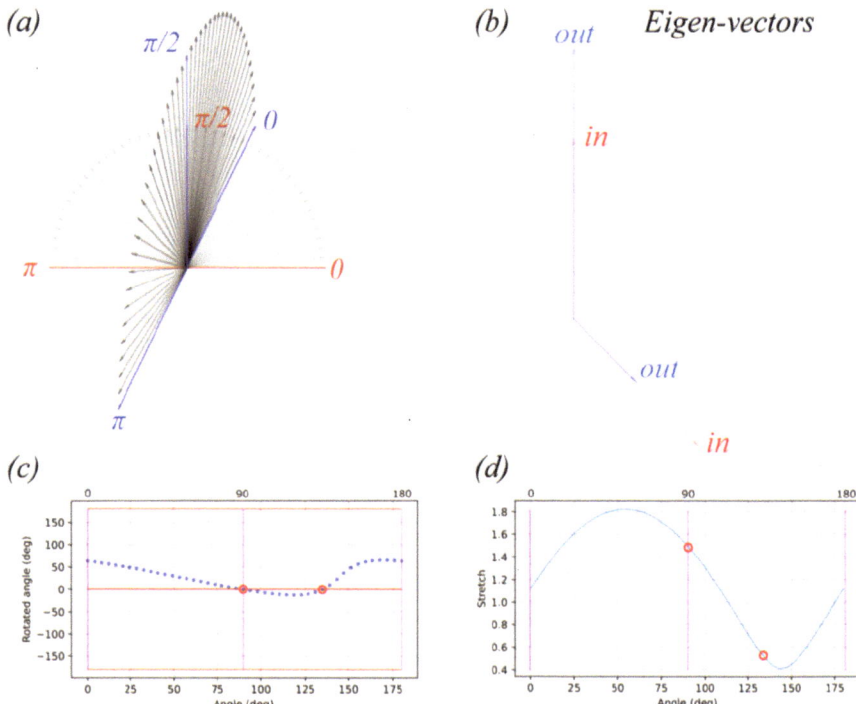

Figure 4.8 Graphical representation of a linear operator $\widehat{L}$ with components: $L_{11} = 5/10, L_{12} = 2, L_{21} = 0, L_{22} = 3/2$. (a) The effect of the operator on unit-length arrows with direction from $0°$ to $180°$; (b) Two special arrow-vectors that do not change their direction under the action of the operator $\widehat{L}$. The operator simply scales the arrows. Such vectors are called *eigen-vectors* of the operator; (c) Horizontal axis: initial orientation of the unit-length arrow; vertical axis: rotation angle of each unit-length arrow. Two special angles are highlighted. For each special angle, the arrow is not rotated. Such arrows are called eigen-vectors of the operator $\widehat{L}$; (d) Horizontal axis: initial orientation of the unit-length arrow; vertical axis: the scaling factor of each unit-length arrow.

4.4 EIGEN-PROBLEM*

Eigen-vectors of a linear operator are special vectors that are not rotated by the operator, eigen-vectors can only be scaled[6]. This requirement can

---

[6]Strictly speaking, it is possible to "flip" the direction of a vector, effectively rotating it by 180 degrees. However, the resulting vector is still parallel to the original one.

be expressed using a simple equation:

$$\widehat{L}\,\vec{a} = \alpha\vec{a},$$

or, if there exists another eigen-vector $\vec{b}$ different from $\vec{a}$:

$$\widehat{L}\,\vec{b} = \beta\vec{b}.$$

Here the numbers α and β specify the scaling coefficients. The equations given above are called *eigen-problem equations*. The vectors $\vec{a}$ and $\vec{b}$ are called *eigen-vector*, the scaling factors α and β are called *eigen-values*. In general, $\alpha \neq \beta$, as illustrated in Figure 4.8(b).

Not all linear operators have eigen-vectors. A simple example is the operator of rotation by a finite angle $\theta \neq 180°$:

$$\widehat{R}_\theta\,\vec{a} \neq \alpha\vec{a}.$$

Simply speaking, it is impossible to rotate a vector without changing its direction.

All "well-behaved" linear operators[7] (excluding rotation operators like $\widehat{R}_\theta$) have the same number of eigen-vectors as the number of basis vectors. Any "well-behaved" linear operator that operates on vectors in a plane has two eigen-vectors, as illustrated in Figure 4.8(b). Operators acting on vectors in three dimensions may have up to three eigen-vectors.

It is possible to find linear operators that have fewer eigen-vectors than the number of basis vectors. Such operators are called *degenerate operators*. They are important and we will consider them next for the special case of two dimensions.

4.5 DEGENERATE LINEAR OPERATORS

In some special cases a linear operator $\widehat{L}$, when applied to the basis arrows $\vec{e}_1$ and $\vec{e}_2$, can produce parallel arrows:

$$(\widehat{L}\,\vec{e}_1) \parallel (\widehat{L}\,\vec{e}_2)$$

or

$$\widehat{L}\,\vec{e}_1 = \lambda(\widehat{L}\,\vec{e}_2)$$

for some number λ.

---

[7]We will encounter ill-behaved linear operators in the next section.

Using components notation, this condition can be written as follows:

$$L_{11}\vec{e}_1 + L_{12}\vec{e}_2 = \lambda L_{21}\vec{e}_1 + \lambda L_{22}\vec{e}_2.$$

The vector on the left-hand side is the same vector as on the right-hand side if they have the same components in the given basis. Therefore, we must equate the corresponding components:

$$L_{11} = \lambda L_{21} \qquad \text{and} \qquad L_{12} = \lambda L_{22}.$$

Cross-multiplying these equations, we obtain

$$\lambda L_{11} L_{22} = \lambda L_{12} L_{21},$$

from which follows

$$L_{12} L_{21} - L_{11} L_{22} = 0.$$

Linear operator satisfying this condition is called *degenerate* for the reasons explained below.

Exercise 4.3 *Show that a degenerate linear operator* $\widehat{L}$ *"collapses" all vectors onto the same line, i.e. all* $(\widehat{L}\,\vec{a})$ *have the same direction.*

Later, in Section 5.7, we will encounter a whole class of useful degenerate linear operators, called *projectors*, whose job is to project any vector onto a specified direction.

4.5.1 Determinant of a Linear Operator

A non-zero vector has length – its simple numeric characteristic independent of the choice of basis. Operators do not have lengths, but they do have certain numeric characteristics that also do not depend on the choice of basis. The two important characteristics of any linear operator are *determinant* and *trace*.

The quantity

$$\boxed{D = L_{12} L_{21} - L_{11} L_{22}}$$

computed from the components of a linear operator is called its *determinant*. Determinant has a clear geometric meaning which can be seen from the action of the operator $\widehat{L}$ on the orthonormal basis, as illustrated in Figure 4.9. Denoting $L_{11} = L_1$, $L_{12} = L_2$, $L_{21} = L_3$, and $L_{22} = L_4$, we can calculate the area of the parallelogram built from the vectors $(\widehat{L}\,\vec{u}_1)$ and $(\widehat{L}\,\vec{u}_2)$ as follows:

$$A = (L_1 + L_3)(L_2 + L_4) - 2(L_1 L_2/2 + L_3 L_4/2 + L_2 L_3).$$

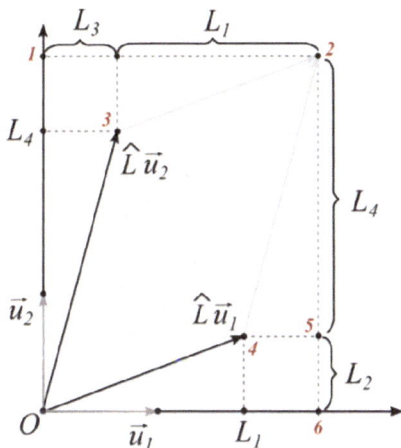

Figure 4.9 A linear operator $\widehat{L}$ transforms basis vectors $\vec{u}_i$ into $\vec{v}_i = \widehat{L}\vec{u}_i$. For non-degenerate operators the vectors $\vec{v}_i$ form sides of a parallelogram with non-zero area $A = L_{12}L_{21} - L_{11}L_{22}$.

It is simply the area of the rectangle $O126$ minus the areas of two rectangles with sides L_2 and L_3, together with two pairs of triangles.

After simplification, the last equation reduces to

$$A = L_1 L_4 - L_2 L_3 = L_{11}L_{22} - L_{12}L_{21}.$$

Thus, the determinant of a linear operator $\widehat{L}$ equals to the ratio of the area of the parallelogram built from the vectors $(\widehat{L}\vec{u}_1)$ and $(\widehat{L}\vec{u}_2)$ to the area of the parallelogram built from the unit vectors $\vec{u}_1$ and $\vec{u}_2$. Put differently, a determinant measures how much an area of the parallelogram built from basis vectors is distorted by an operator. Obviously, for a degenerate linear operator, this ratio is zero, since all arrows "collapse" onto a single direction.

Determinant

For arrow-vectors in three and higher dimensions, we also have linear operators. The meaning of determinant in these higher-dimensional cases remains similar: Determinant expresses the ratio of *volumes* built from basis vectors $\vec{e}_1, \vec{e}_2 \ldots, \vec{e}_n$, and from their "transformed" versions $\widehat{L}\vec{e}_1, \widehat{L}\vec{e}_2 \ldots, \widehat{L}\vec{e}_n$.

The second numeric characteristic of a linear operator – called *trace* – will be explored below in the Exercise 4.5, after we learn how the components L_{ij} of a linear operator $\widehat{L}$ transform when we change basis.

4.6 USING OPERATOR COMPONENTS

It is possible, and is often convenient, to work with a linear operator using only its components, without referring to arrows or some other graphical representation.

The action of a linear operator $\widehat{L}$ on a vector $\vec{a}$ can be written in terms of components:

$$\widehat{L}\,\vec{a} = \widehat{L}\,(a_i\vec{e}_i) = a_i\,(\widehat{L}\,\vec{e}_i) = a_i(L_{ij}\vec{e}_j).$$

On the other hand

$$\widehat{L}\,\vec{a} = \vec{b} = b_j\vec{e}_j.$$

Comparing the last two expressions, we observe that the action of a linear operator can be written entirely in components:

$$\boxed{a_i L_{ij} = b_j.}$$

Exercise 4.4 *Show that the same relation holds in any other basis. Namely, prove that*

$$a_i' L_{ij}' = b_j'.$$

The relation between components of the operator $\widehat{L}$ and vectors $\vec{a}$ and $\vec{b}$ can be written differently:

$$b_j = L_{ij}a_i \quad \rightarrow \quad b_j = L_{ji}^T a_i \quad \rightarrow \quad b_i = L_{ij}^T a_j.$$

Here we used the operation of transposition (introduced in the subsection 3.4.1 on page 55) and index renaming to make summation over the second index of L_{ij}^T.

4.7 COMPONENTS TRANSFORMATION

Warning: This section has the highest density of algebraic manipulations in the whole book. However, the manipulations are rather trivial, they consist only in multiplications and additions. The challenge for the reader is to stay focused and follow the derivations closely because the final result is very important. The symbolic operations in this section are typical for linear algebra. Without good notation, such calculations may become too tedious. This is why we will first demonstrate how to use Einstein's summation rule to quickly get the desired result. Only after that we will derive the same result again, showing all steps in detail. Now let's get the job done!

At the first look at components, the difference between vectors and linear operators is in the number of component indices: one for vectors (a_i), and two for linear operators (L_{ij}). To further compare vectors and linear operators, we can study how the components of the latter transform between different bases.

Power of ESR and Index Notation

We first demonstrate the power of Einstein's summation rule and index notation and find how the transformation relations can be quickly deduced. We start with the expansion of an operator in the "new" (primed) basis:

$$\widehat{L}\,\vec{e}\,'_i = L'_{il}\,\vec{e}\,'_l. \tag{4.2}$$

Next, we use the relation between "new" and "old" basis vectors and the linearity of $\widehat{L}$ to fully expand the left-hand side:

$$\vec{e}\,'_i = E_{ij}\vec{e}_j, \quad \vec{e}_k = E'_{kl}\vec{e}\,'_l,$$

$$\widehat{L}\,\vec{e}\,'_i = \widehat{L}\,(E_{ij}\vec{e}_j) = E_{ij}(\widehat{L}\,\vec{e}_j) = E_{ij}L_{jk}\vec{e}_k = E_{ij}L_{jk}E'_{kl}\vec{e}\,'_l.$$

Comparing this to the right-hand side of the equation (4.2), we arrive at

$$\boxed{L'_{il} = E_{ij}L_{jk}E'_{kl}.}$$

Now we will repeat the steps in detail.

The components (L'_{ij}) of a linear operator $\widehat{L}$ in the "new" (primed) basis are determined in the same way as for the "old" basis:

$$\widehat{L}\,\vec{e}\,'_1 = L'_{11}\vec{e}\,'_1 + L'_{12}\vec{e}\,'_2,$$

$$\widehat{L}\,\vec{e}\,'_2 = L'_{21}\vec{e}\,'_1 + L'_{22}\vec{e}\,'_2.$$

Next, in the left-hand sides of these equations, we replace $\vec{e}\,'_i$ with their expansion in the "old" basis and use the linearity of the operator $\widehat{L}$:

$$\widehat{L}\,\vec{e}\,'_1 = \widehat{L}\,(E_{11}\vec{e}_1 + E_{12}\vec{e}_2) = E_{11}(\widehat{L}\,\vec{e}_1) + E_{12}(\widehat{L}\,\vec{e}_2)$$

and
$$\widehat{L}\,\vec{e}\,'_2 = \widehat{L}\,(E_{21}\vec{e}_1 + E_{22}\vec{e}_2) = E_{21}(\widehat{L}\,\vec{e}_1) + E_{22}(\widehat{L}\,\vec{e}_2).$$

The action of $\widehat{L}$ on the "old" basis is determined by the components L_{ij}:

$$\widehat{L}\,\vec{e}_i = L_{ij}\vec{e}_j.$$

Using these relations, we further transform

$$\widehat{L}\,\vec{e}\,'_1 = E_{11}(L_{11}\vec{e}_1 + L_{12}\vec{e}_2) + E_{12}(L_{21}\vec{e}_1 + L_{22}\vec{e}_2)$$

and

$$\widehat{L}\,\vec{e}\,'_2 = E_{21}(L_{11}\vec{e}_1 + L_{12}\vec{e}_2) + E_{22}(L_{21}\vec{e}_1 + L_{22}\vec{e}_2).$$

Opening the parentheses and grouping the terms with identical basis vectors, we get

$$\widehat{L}\,\vec{e}\,'_1 = \mu_1\vec{e}_1 + \mu_2\vec{e}_2$$

and

$$\widehat{L}\,\vec{e}\,'_2 = \nu_1\vec{e}_1 + \nu_2\vec{e}_2,$$

where, in order to avoid clutter, we introduced notation

$$\mu_1 = E_{11}L_{11} + E_{12}L_{21}, \quad \mu_2 = E_{11}L_{12} + E_{12}L_{22}$$

and

$$\nu_1 = E_{21}L_{11} + E_{22}L_{21}, \quad \nu_2 = E_{21}L_{12} + E_{22}L_{22}.$$

Finally, we expand the "old" basis vectors in terms of the "new" ones, to arrive at

$$\widehat{L}\,\vec{e}\,'_1 = \mu_1(E'_{11}\vec{e}\,'_1 + E'_{12}\vec{e}\,'_2) + \mu_2(E'_{21}\vec{e}\,'_1 + E'_{22}\vec{e}\,'_2)$$

and

$$\widehat{L}\,\vec{e}\,'_2 = \nu_1(E'_{11}\vec{e}\,'_1 + E'_{12}\vec{e}\,'_2) + \nu_2(E'_{21}\vec{e}\,'_1 + E'_{22}\vec{e}\,'_2).$$

Opening the parentheses and grouping the terms with identical basis vectors results in

$$\widehat{L}\,\vec{e}\,'_1 = (\mu_1 E'_{11} + \mu_2 E'_{21})\vec{e}\,'_1 + (\mu_1 E'_{12} + \mu_2 E'_{22})\vec{e}\,'_2$$

and

$$\widehat{L}\,\vec{e}\,'_2 = (\nu_1 E'_{11} + \nu_2 E'_{21})\vec{e}\,'_1 + (\nu_1 E'_{12} + \nu_2 E'_{22})\vec{e}\,'_2.$$

From the last two equations, we can read the components L'_{ij} of the operator $\widehat{L}$ in terms of its components L_{mn}:

$$L'_{11} = \mu_1 E'_{11} + \mu_2 E'_{21}, \quad L'_{12} = \mu_1 E'_{12} + \mu_2 E'_{22},$$

$$L'_{21} = \nu_1 E'_{11} + \nu_2 E'_{21}, \quad L'_{22} = \nu_1 E'_{12} + \nu_2 E'_{22}.$$

To keep the formulas manageable, we will use the summation convention and write first

$$L'_{11} = \mu_i E'_{i1}, \quad L'_{12} = \mu_i E'_{i2},$$

$$L'_{21} = \nu_i E'_{i1}, \quad L'_{22} = \nu_i E'_{i2}.$$

Then, observing the matching indices, we can shorten these relations even further:

$$L'_{1j} = \mu_i E'_{ij},$$

$$L'_{2j} = \nu_i E'_{ij}.$$

After that, having looked at the expressions for μ_i and ν_i, we can shorten them to the following:

$$\mu_i = E_{1k} L_{ki} \text{ and } \nu_i = E_{2k} L_{ki}.$$

Plugging these expressions into the formulas for L'_{ij}, we obtain

$$L'_{1j} = E_{1k} L_{ki} E'_{ij} \text{ and } L'_{2j} = E_{2k} L_{ki} E'_{ij}.$$

Finally, comparing the indices, we arrive at the most compact form of relations:

$$L'_{mj} = E_{mk} L_{ki} E'_{ij}. \tag{4.3}$$

Comparing this to the transformation of components of a contravariant vector:

$$a'_j = a_i E'_{ij},$$

we see both similarities and differences.

The similarity is seen in the transformation rule for the second index of the operator $\widehat{L}$: It transforms according to the *contravariant* rule. The difference is seen in the transformation rule for the first index which transforms according to the *covariant* rule.

We conclude that linear operators $\widehat{L}$ which map vector into vectors as

$$\vec{a} \xrightarrow{\widehat{L}} \vec{b}$$

have a "mixed nature": They combine the behavior of covariant and contravariant vectors, as seen from the transformation of their components. Soon we will encounter linear operators whose components transform like two covariant vectors, or like two contravariant vectors (see *tensor product* in Section 5.8).

Objects and Components

Notions of vectors and operators are *independent of any basis and components*. Components simply *represent* vectors or operators in a given basis. Neither vectors nor operators are reduced to their components, like a building is not reduced to its projections on a piece of paper.

Components are simply computational tools, but their transformation properties tell something about the corresponding mathematical object.

Exercise 4.5 *The area of the parallelogram built on unit vectors is 1. The area of the parallelogram built on the vectors $(\widehat{L}\,\vec{u}_1)$ and $(\widehat{L}\,\vec{u}_2)$ does not depend on the basis used to express the components of the operator $\widehat{L}$. Therefore, the determinant should be a quantity independent of the basis.*

Another characteristic of an operator which is independent of the basis is called trace. *It is defined as the sum of the diagonal components of an operator:*

$$tr\,\widehat{L} = L_{11} + L_{22}.$$

Prove that trace is independent of the choice of a specific basis. That is, show that, given two sets of basis vectors $\{\vec{e}_i\}$ and $\{\vec{e}\,'_i\}$, the following equality holds

$$L_{11} + L_{22} = L'_{11} + L'_{22}.$$

4.8   FIRST NOTION OF TENSOR

Although a more satisfactory definition of tensors will be deduced later, we can now recognize at least one type of tensor. This type of tensor is represented by linear operators considered above.

Tensors are mathematical objects which, like numbers and vectors, can be added pairwise, multiplied by numbers, and, very importantly, *whose components transform in a certain way.* The components of linear operators of the type $\widehat{L}\,\vec{a} = \vec{b}$ transform according to the formula

$$L'_{ij} = E_{im} L_{mn} E'_{nj}.$$

In Section 5.8 we will encounter tensors with other transformation rules. However, those rules will be analogous to the expression given above. We will also learn what it means to add two tensors, and even what it means to "multiply" a pair of tensors.

Finally, another important aspect of operators and tensors can be seen when we write

$$\vec{b} = \widehat{L}\,\vec{a} \quad \text{or} \quad b_j = a_i L_{ij}.$$

These formulas express a *linear relation* between two vectors, and the operator $\widehat{L}$ (or tensor) encodes that relation. Linear relations between vector quantities are very important in physics. Examples of this are given in the last chapter.

CHAPTER HIGHLIGHTS

- *Operators extend the idea of functions.*

- *Numeric functions (e.g., $\sin x$) act on numbers and yield other numbers. Operators may act on vectors to yield other vectors or numbers.*

- *Linear operators represent a simple and yet powerful class of operators on vectors.*

- *Linear operators can be represented graphically or symbolically.*

- *In a given basis, every linear operator can be specified using components. This is similar to how a vector is represented via its components.*

- *When the basis changes, the components of a linear operator transform in a very specific way, called transformation rule or law.*

- *Mathematical expressions involving linear operators can be written using components (e.g., L_{ij}) or they can be written in more abstract operator form using operator notation $\widehat{L}$.*

- *Degenerate linear operators collapse two or more basis vectors onto the same line. They form a special subset of linear operators.*

- *There are two characteristics of linear operators that do not change when a new basis is chosen: Determinant and trace.*

- *Linear operators represent the simplest type of tensors, called tensors of the second rank – one tier above vectors.*

- *Linear operators and tensors are used to express linear relations between different vector quantities.*

CHAPTER 5

Tensors

T HE FIRST EXAMPLE OF AN OPERATOR – AND corresponding tensor – that we encountered had a simple type:

$$\widehat{L}\,\vec{a} = \vec{b}.$$

It is a linear unary function mapping vectors into vectors.

As the next step, we will expand our toolset and study operators that take two vectors as their input arguments. Their action on the arguments can be symbolically written as follows:

$$\widehat{L}\,\vec{a}\,\vec{b} = x$$

if the result is a number, or

$$\widehat{L}\,\vec{a}\,\vec{b} = \vec{c}$$

if the result is another vector. We will start with the former case.

5.1   SCALAR PRODUCT AND DOL-OPERATOR

Given a pair of vectors, for example, arrows in a plane, we can *quantify their mutual orientation*. In other words, given two vectors $\vec{a}$ and $\vec{b}$, we can calculate a number that measures, for instance, their *degree of overlap* (or *alignment*) which tells how much in common the vectors have with regard to their directions and lengths.

We are looking for a binary operator $\widehat{\sigma}$ that yields a number based on two vectors:

$$\widehat{\sigma}\,\vec{a}\,\vec{b} = x.$$

DOI: 10.1201/9781003620365-5

We will call this operator $\hat{\sigma}$ *dol*-operator[1], based on the key letters of the phrase "degree of overlap."

There are many approaches to quantify mutual orientation of a pair of vectors. One simple way is to measure the angle between them:

$$\angle \, \vec{a} \, \vec{b} = \theta.$$

However, we will limit our search for candidates to linear operators or – in the case of binary operator $\hat{\sigma}$ – to *bilinear operators*.

Exercise 5.1 *Prove that the operator $\angle$ is not linear.*

If the dol-operator $\hat{\sigma}$ is linear in each of its arguments, we must have

$$\hat{\sigma}\,(2\vec{a})\,\vec{b} = \hat{\sigma}\,(\vec{a} + \vec{a})\,\vec{b} = \hat{\sigma}\,\vec{a}\,\vec{b} + \hat{\sigma}\,\vec{a}\,\vec{b} = 2(\hat{\sigma}\,\vec{a}\,\vec{b}).$$

In general, we require

$$\hat{\sigma}\,(\alpha\vec{a})\,\vec{b} = \alpha(\hat{\sigma}\,\vec{a}\,\vec{b}).$$

In addition, we will not be concerned with the order in which the vectors $\vec{a}$ and $\vec{b}$ are considered. Put differently, we consider $\vec{a}$ aligned to $\vec{b}$ to the same degree as $\vec{b}$ is aligned to $\vec{a}$. Mathematically this requirement is written as follows:

$$\hat{\sigma}\,\vec{a}\,\vec{b} = \hat{\sigma}\,\vec{b}\,\vec{a}.$$

Functions and operators with this property are called *symmetric* in their arguments.

Anti-symmetric Operators

Since angles are measured as clockwise (negative) or counter-clockwise (positive), the operator $\angle$ is not symmetric. It is called *anti-symmetric*:

$$\angle \, \vec{a} \, \vec{b} = -(\angle \, \vec{b} \, \vec{a}).$$

---

[1]This is not a standard terminology, but it conveys the concept well.

Finally, we recognize that some pairs of vectors are not overlapping at all. It can be said that their degree of overlap is zero. Mutually orthogonal vectors provide an example of vectors with zero overlap, they kind of "have nothing in common." Thus, we expect

$$\widehat{\sigma}\, \vec{a}\, \vec{b} = 0$$

if $\vec{a} \perp \vec{b}$.

Now every vector can be written in the form

$$\vec{a} = a\vec{u}_a,$$

where a is the length of the vector, $\vec{u}_a$ is the unit-length vector pointing in the same direction as $\vec{a}$. Similarly for $\vec{b}$:

$$\vec{b} = b\vec{u}_b.$$

Since dol-operator $\widehat{\sigma}$ is linear in each argument, we can write

$$\widehat{\sigma}\, \vec{a}\, \vec{b} = ab(\widehat{\sigma}\, \vec{u}_a\, \vec{u}_b).$$

The problem is thus reduced to quantifying the mutual overlap of unit vectors $\vec{u}_a$ and $\vec{u}_b$. The only difference between these vectors is their directions. Furthermore, only relative angle should matter, since rotating both vectors by the same amount should not affect their mutual/relative orientation. Although we can't use the angle between these vectors directly, we can use some function of the angle:

$$\widehat{\sigma}\, \vec{u}_a\, \vec{u}_b = f\,\theta.$$

Firstly, this function has to be periodic since changing the mutual angle by 2π results in the same mutual orientation. Secondly, for orthogonal vectors the function must return zero degree of overlap:

$$f\left(\pi/2\right) = 0.$$

After some search and reflection, a reasonable candidate function can be written in the following way:

$$f\,\theta = \cos\theta.$$

We arrived at the traditional operation of *scalar product* (numeric product) of two vectors. For scalar product, a special *infix notation* is conventionally used:

$$\hat{\sigma}\,\vec{a}\,\vec{b} = \vec{a} \cdot \vec{b} = ab\cos\theta.$$

What is the geometric meaning of scalar product? As shown in the Figure 5.1, one way to interpret the scalar product $\vec{a} \cdot \vec{b} = ab\cos\theta$ is to consider it as the area of a rectangle with sides a and $b\cos\theta$. A special care must be taken for cases when the mutual angle is greater than π.

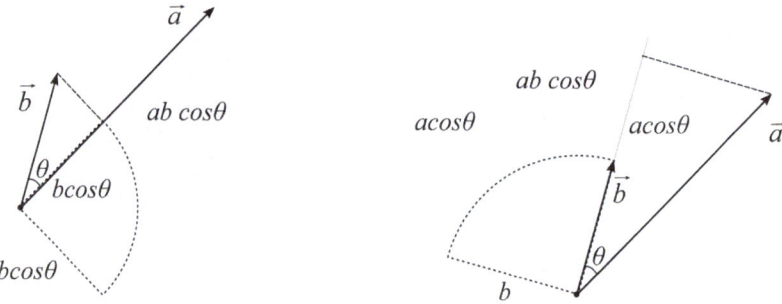

Figure 5.1   Scalar product of two vectors can be given a simple geometric interpretation as the unsigned area of a rectangle with the sides a and $b\cos\theta$ or with the sides $a\cos\theta$ and b.

5.2   SCALAR PRODUCT PROPERTIES

The expression for the scalar product of two vectors

$$\vec{a} \cdot \vec{b} = ab\cos\theta$$

implies the *commutativity* of scalar product operation:

$$\vec{b} \cdot \vec{a} = ba\cos\theta = ab\cos\theta = \vec{a} \cdot \vec{b}.$$

Non-commutativity

Commutativity is a familiar property, present in both addition and multiplication of numbers. It is sometimes accepted as a natural property of any multiplication-like operation. This view is limiting, however, and we will later learn how to multiply objects without commutativity:

$$A \bowtie B \neq B \bowtie A.$$

Here we used an arbitrary infix symbol ⋈ to denote the non-commutative product of some objects A and B. What exactly those objects are and what their product might mean will be clear when we introduce them properly. Right now we want to highlight the non-commutativity as a valid property of some binary operations.

Besides commutativity, scalar product has other useful properties. First, it is trivial to demonstrate that we can "pull out" constant scale factors of vectors:

$$(\alpha \vec{a}) \cdot \vec{b} = \alpha(\vec{a} \cdot \vec{b})$$

and similarly

$$\vec{a} \cdot (\beta \vec{b}) = \beta(\vec{a} \cdot \vec{b}).$$

To arrive at the second important property of the scalar product, recall that in the geometric interpretation of the scalar product we had expressions

$$b \cos \theta = b_{\parallel} \quad \text{or} \quad a \cos \theta = a_{\parallel},$$

where $b_{\parallel}$ is the part of the vector $\vec{b}$ parallel to $\vec{a}$ ($a_{\parallel}$ is the part of the vector $\vec{a}$ parallel to $\vec{b}$.)

Now, if the vector $\vec{a}$ is "made from" two other vectors $\vec{c}$ and $\vec{d}$:

$$\vec{a} = \vec{c} + \vec{d},$$

then, as illustrated in the Figure (5.2),

$$a_{\parallel} = c_{\parallel} + d_{\parallel},$$

where all terms represent parts of the respective vectors parallel to the vector $\vec{b}$. Of course, the angles between the vectors $\vec{a}$, $\vec{c}$, $\vec{d}$ and the vector $\vec{b}$ may all be different:

$$a_{\parallel} = a \cos \theta = c \cos \phi + d \cos \psi.$$

Given that

$$\vec{a} \cdot \vec{b} = ab \cos \theta, \quad \vec{c} \cdot \vec{b} = cb \cos \phi, \quad \vec{d} \cdot \vec{b} = db \cos \psi,$$

we arrive at the *distributive property* of scalar product:

$$(\vec{c} + \vec{d}) \cdot \vec{b} = \vec{a} \cdot \vec{b} = (\vec{c} \cdot \vec{b}) + (\vec{d} \cdot \vec{b}).$$

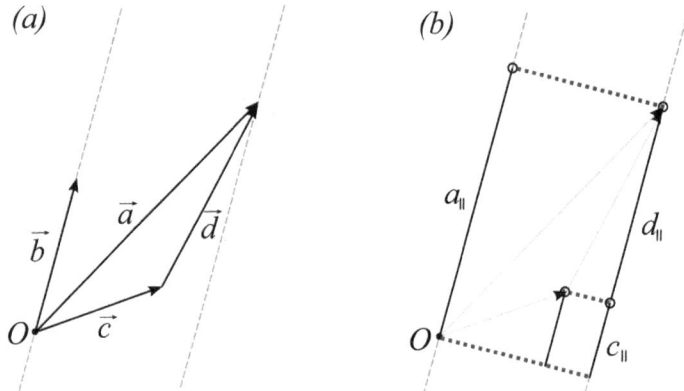

Figure 5.2 Simple geometric construction illustrates the distributive property of scalar product: $(\vec{c} + \vec{d}) \cdot \vec{b} = (\vec{c} \cdot \vec{b}) + (\vec{d} \cdot \vec{b})$.

Putting it all together, we can express the properties of scalar product in a single expression:

$$(\alpha \vec{a} + \beta \vec{b}) \cdot \vec{c} = \vec{c} \cdot (\alpha \vec{a} + \beta \vec{b}) = \alpha(\vec{a} \cdot \vec{c}) + \beta(\vec{b} \cdot \vec{c}).$$

Orthogonality

We might wonder whether we can define, instead of the degree of overlap, some measure of orthogonality for a pair of vectors. Indeed, a reasonable candidate might be (using infix notation):

$$\vec{a} \vdash \vec{b} = ab \sin \theta.$$

While this is an acceptable definition, the expression on the right-hand side appears in a different, more powerful, and useful, operation of the *outer product* of two vectors. The concept of outer product is related to the concept of *tensor product* (see Section 5.8) but is beyond the scope of this book.

Using the properties of the scalar product and expanding vectors in terms of the basis vectors, we can write the scalar product in terms of vector components. First write

$$(a_1 \vec{e}_1 + a_2 \vec{e}_2) \cdot \vec{b} = a_1(\vec{b} \cdot \vec{e}_1) + a_2(\vec{b} \cdot \vec{e}_2).$$

Next, expand $\vec{b}$ to find

$$(b_1\vec{e}_1 + b_2\vec{e}_2) \cdot \vec{e}_i = b_1(\vec{e}_1 \cdot \vec{e}_i) + b_2(\vec{e}_2 \cdot \vec{e}_i).$$

Finally, combining the results of two previous steps, we arrive at the expression

$$\vec{a} \cdot \vec{b} = a_1b_1(\vec{e}_1 \cdot \vec{e}_1) + a_1b_2(\vec{e}_2 \cdot \vec{e}_1) + a_2b_1(\vec{e}_1 \cdot \vec{e}_2) + a_2b_2(\vec{e}_2 \cdot \vec{e}_2).$$

For arbitrary basis vectors the scalar product is then given by

$$\boxed{\vec{a} \cdot \vec{b} = a_1b_1e_1^2 + (a_1b_2 + a_2b_1)e_1e_2\cos\theta + a_2b_2e_2^2,}$$

where θ is the angle between the basis vectors $\vec{e}_1$ and $\vec{e}_2$. For a special case of orthonormal basis the scalar product takes the simplest form:

$$\boxed{\vec{a} \cdot \vec{b} = a_1b_1 + a_2b_2 = a_ib_i.}$$

But in general we must know the values for all products $\vec{e}_i \cdot \vec{e}_j$. A special notation is used for these products:

$$\eta_{ij} = \vec{e}_i \cdot \vec{e}_j = \hat{\sigma}\,\vec{e}_i\,\vec{e}_j.$$

This notation allows a more compact way of writing scalar products for general basis:

$$\boxed{\vec{a} \cdot \vec{b} = \eta_{ij}a_ib_j.}$$

Scalar Product Components

Let's take a look, how the derivation of the last result can be done using index notation.

$$\vec{a} \cdot \vec{b} = (a_i\vec{e}_i) \cdot \vec{b} = a_i(\vec{e}_i \cdot \vec{b}).$$

Expanding $\vec{b}$, we get

$$\vec{a} \cdot \vec{b} = a_i(\vec{e}_i \cdot [b_j\vec{e}_j]) = a_i(b_j[\vec{e}_i \cdot \vec{e}_j]).$$

We thus showed that

$$\vec{a} \cdot \vec{b} = a_ib_j\vec{e}_i \cdot \vec{e}_j.$$

In the process, we had to use twice the distributive property of the scalar product, as well as the distributive property of number multiplication.

Exercise 5.2 *In the index form of the scalar product of two vectors*

$$\vec{a} \cdot \vec{b} = (a_i b_j)(\vec{e}_i \cdot \vec{e}_j)$$

we observe the expression with two indices:

$$\beta_{ij} = a_i b_j.$$

Can β_{ij} represent the components of some linear operator $\widehat{\beta}$? If so, how does this operator act on vectors?

5.3   INNER OPERATIONS

The following fact is easy to take for granted and overlook: *Vectors are not used by themselves. They need numbers.* Without multiplying a vector by a number, we could not write the simplest expansion of a vector $\vec{a}$ in a basis:

$$\vec{a} = a_1 \vec{e}_1 + a_2 \vec{e}_2.$$

Note that all operations we considered so far never took us outside of the combined realm of numbers $\mathbb{R}$ and vectors $\overset{\Rightarrow}{A}$. Indeed, a product of a number and a vector $\alpha \vec{a}$ uses one element of each space and produces a vector. Similarly, a sum of two vectors $\vec{a} + \vec{b}$ takes two elements from $\overset{\Rightarrow}{A}$ and returns another element of $\overset{\Rightarrow}{A}$. Finally, the dol-operator $\widehat{\sigma}$ takes two elements from $\overset{\Rightarrow}{A}$ and returns an element of $\mathbb{R}$. These points are illustrated in Figure 5.3(a).

Another way of looking at this reveals that in the hierarchy of mathematical objects, the operations we considered so far never took us up the ladder of *ranks*, as illustrated in Figure 5.3(b). At the lowest level, we have *rank-0* elements – numbers. The first ladder corresponds to the *rank-1* elements – vectors. One step higher we have *rank-2* elements – tensors of the second rank, and so on.

All operations that *do not* result in the element of higher rank are called *inner operation*. For instance, the scalar product of vectors is often called *inner product*[2]. The phrase *inner sum* is not used for the binary operator $(+)$.

---

[2]There also exists an *outer product*, which is related to the *tensor product* discussed in Section 5.8.

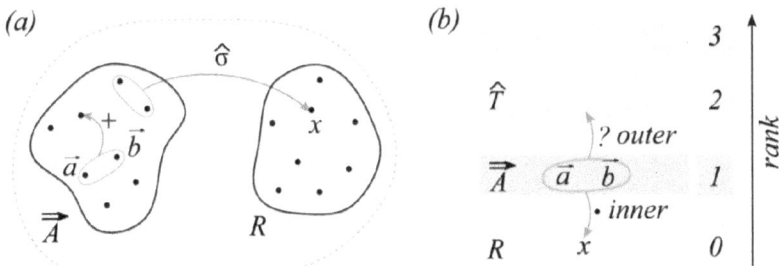

Figure 5.3 (a) In vector algebra we use number space $\mathbb{R}$ alongside vector space $\overset{\Rightarrow}{A}$. All operations considered so far yield results from either of these spaces; such operations (sum, product, etc.) are called *inner operations*. (b) In contrast to inner operations, *outer operations* create mathematical objects of higher rank than the input arguments. For example, the outer product of two vectors yields a second-rank tensor.

5.4 CONJUGATE OBJECTS

Starting from the idea of scalar product or, equivalently, from the dol-operator

$$\widehat{\sigma}\,\vec{a}\,\vec{b} = x$$

we can arrive at an important notion of *conjugate* objects. Two objects are conjugate of each other if, roughly speaking, they are somehow related via a simple rule. We will study this notion using vectors.

For every vector $\vec{a}$ there exists a mathematical object, related to $\vec{a}$ via the dol-operator. To see this, we first need to revisit the idea of *partial application*, discussed in subsection 5.4.1. This time, however, we will extend the idea of partial application to binary operators.

5.4.1 Partial Application

Given a pair of vectors, the dol-operator yields a number:

$$\widehat{\sigma}\,\vec{a}\,\vec{b} = x.$$

What happens when we provide only one vector, leaving the second-input slot of $\widehat{\sigma}$ empty (the box here indicates a missing second argument):

$$\widehat{\sigma}\,\vec{a}\,\square\,?$$

This is called a *partial application* of the operator $\hat{\sigma}$. This construction has the behavior of a unary operator that maps any vector into a number:

$$\vec{c} \xrightarrow{\hat{\sigma}\,\vec{a}} y.$$

We will use a special notation for the partially applied operator:

$$\boxed{\overleftarrow{a} = \hat{\sigma}\,\vec{a}.}$$

Linear Operator

The unary operator $\overleftarrow{a}$ is a *linear operator*:

$$\overleftarrow{a}\,(\vec{b} + \vec{c}) = (\overleftarrow{a}\,\vec{b}) + (\overleftarrow{a}\,\vec{c}).$$

The action of a unary operator $\overleftarrow{a}$ on any vector is then naturally defined as

$$\overleftarrow{a}\,\vec{b} = \hat{\sigma}\,\vec{a}\,\vec{b} = ab\cos\theta.$$

Notice that in the left-most expression the operator $\overleftarrow{a}$ and the argument vector $\vec{b}$ are separated by space, in agreement with the notation for the application of functions and operator to their arguments.

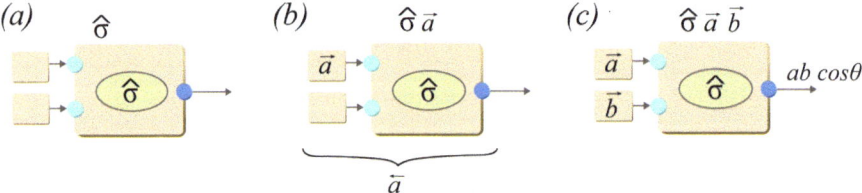

Figure 5.4 (a) The dol-operator is a binary operator linear in each of its arguments (*bilinear*). (b) When only one argument is supplied, the operator becomes *partially applied* and is denoted using the *conjugate vector* notation $\overleftarrow{a}$. (c) Applied to two vector arguments (*fully applied*), dol-operator yields a number – scalar product of the vectors.

All three possible "states" of the dol-operator $\hat{\sigma}$ are illustrated in Figure 5.4. The middle state of partial application corresponds to a linear operator built from the first input vector $\vec{a}$ and as denoted as $\overleftarrow{a}$. The operator $\overleftarrow{a}$ is called the *conjugate* to a vector $\overrightarrow{a}$. The conjugation is understood *relative to the scalar product or, equivalently, binary operator* $\widehat{\sigma}$.

Quantum Notation

In quantum physics vectors and their duals are used to describe information about quantum systems (so called quantum states.) Paul Dirac introduced a very powerful notation for these vectors, called *bra* and *ket* vectors. Ket vectors correspond to contravariant arrow-vectors and are denoted as follows:

$$\overrightarrow{a} \quad \longrightarrow \quad |a\rangle.$$

Bra vectors correspond to covariant vectors and are denoted as

$$\overleftarrow{a} \quad \longrightarrow \quad \langle a|.$$

Scalar product of bra and ket vectors is then written using *brackets*:

$$\overleftarrow{a}\,\overrightarrow{a} = \langle a|a\rangle.$$

5.5 CONJUGATE VECTORS

The notation for the conjugate vectors is suggestive for a reason. The left-pointing arrow on top of $\overleftarrow{a}$ indicates that this *is a vector*. This might be surprising since we just convinced ourselves that $\overleftarrow{a}$ is a linear operator. To convince ourselves that $\overleftarrow{a}$ is also a vector, we must check whether the conjugate of vectors possesses the defining properties of vectors. Let's do it.

To demonstrate that we can add two conjugate vectors – which are also linear operators – we must describe how to add operators. Operators are essentially functions, and we already understand how to add functions of a numerical arguments (go back to subsection 2.3.4 if you need a refresher.) We can add two operators, $\overleftarrow{a}$ and $\overleftarrow{b}$, in a similar way:

$$\overleftarrow{a} + \overleftarrow{b} = \overleftarrow{c},$$

where for any contravariant vector $\overrightarrow{d}$ the following holds:

$$\overleftarrow{c}\,\overrightarrow{d} = (\overleftarrow{a} + \overleftarrow{b})\overrightarrow{d} = (\overleftarrow{a}\,\overrightarrow{d}) + (\overleftarrow{b}\,\overrightarrow{d}).$$

As a side-note, we point out once more that the addition operation "+" in $\overleftarrow{a} + \overleftarrow{b}$ is applied to a new type of mathematical object – conjugate vectors. The addition operator in $\overleftrightarrow{a}d + \overleftrightarrow{b}d$ is applied to usual numbers.

It is easy to see how conjugate vectors can be multiplied by numbers:

$$\alpha\overleftarrow{a} = \overleftarrow{b},$$

where for any contravariant vector $\overrightarrow{c}$ we have

$$\overleftarrow{b}\overrightarrow{c} = \alpha(\overleftrightarrow{a}\overrightarrow{c}).$$

Conjugate Basis

If conjugate objects are vectors, they must have some basis. The basis for conjugate vectors can be taken by conjugating any basis from the arrow-vectors:

$$\boxed{\overleftarrow{e}_1 = \widehat{\sigma}\,\overrightarrow{e}_1, \quad \overleftarrow{e}_2 = \widehat{\sigma}\,\overrightarrow{e}_2.}$$

The linearity of the operator $\widehat{\sigma}$ for both arguments results in the following relation

$$\overleftarrow{a} = \widehat{\sigma}\,\overrightarrow{a} = \widehat{\sigma}\,(a_1\overrightarrow{e}_1 + a_2\overrightarrow{e}_2) = a_1\overleftarrow{e}_1 + a_2\overleftarrow{e}_2.$$

In other words, *any conjugate vector can be expanded in terms of some basis conjugate vectors* $\{\overleftarrow{e}_i\}$.

Exercise 5.3 *Derive the relationship*

$$\overleftarrow{a} = a_1\overleftarrow{e}_1 + a_2\overleftarrow{e}_2$$

in more details, without leaving out steps.

Component Transformation

Finally, we must show that the components of any conjugate vector change properly when the (conjugate) basis is switched.

We start by writing the same operator $\overleftarrow{a}$ in different bases:

$$\overleftarrow{a} = a_1\overleftarrow{e}_1 + a_2\overleftarrow{e}_2 = a_1'\overleftarrow{e}_1' + a_2'\overleftarrow{e}_2'.$$

Here

$$\overleftarrow{e}_1' = \widehat{\sigma}\,\overrightarrow{e}_1', \quad \overleftarrow{e}_2' = \widehat{\sigma}\,\overrightarrow{e}_2'.$$

Expanding the primed basis in terms of the non-primed, and using the linearity of the operator $\widehat{\sigma}$ for all arguments, we get

$$\overset{\leftarrow}{e}{}'_1 = \widehat{\sigma}\,\vec{e}{}'_1 = E_{11}\overset{\leftarrow}{e}_1 + E_{12}\overset{\leftarrow}{e}_2,$$

$$\overset{\leftarrow}{e}{}'_2 = \widehat{\sigma}\,\vec{e}{}'_2 = E_{21}\overset{\leftarrow}{e}_1 + E_{22}\overset{\leftarrow}{e}_2.$$

Plugging these equations into the expansion of $\overset{\leftarrow}{a}$, after grouping the terms, we arrive at

$$\overset{\leftarrow}{a} = a_1\overset{\leftarrow}{e}_1 + a_2\overset{\leftarrow}{e}_2 = (a'_1 E_{11} + a'_2 E_{21})\overset{\leftarrow}{e}_1 + (a'_1 E_{12} + a'_2 E_{22})\overset{\leftarrow}{e}_2.$$

Comparing the coefficients in front of the basis vectors, we conclude that

$$a_1 = a'_i E_{i1},$$

$$a_2 = a'_i E_{i2}.$$

In a more compact form:

$$\boxed{a_j = a'_i E_{ij}.}$$

This is the same form we obtained before for arrow-vectors, with the only (unessential) difference of notation – using E instead of L to denote the relation between the "old" and "new" bases.

The conclusion is as follows: *The conjugate vectors have completely analogous properties as the arrow-vectors, when referred to their own conjugate bases.*

We thus showed that vectors in plane have "conjugate image" – a set of linear operators which also behave like vectors. These conjugate vectors belong to the special vector space called *conjugate vector space* or *dual vector space*. The "usual" vector space is denoted as $\vec{A}$ and its dual companion is denoted as $\overset{\leftarrow}{A}$. The relationship between the "original" and the *conjugate/dual* space is illustrated in the Figure 5.5.

5.6   OPERATORS ARE VECTORS

Starting with the scalar product or, equivalently, bilinear dol-operator $\widehat{\sigma}$, we obtained unary linear operators using partial application

$$\vec{a} \quad \Longrightarrow \quad \overset{\leftarrow}{a} = \widehat{\sigma}\,\vec{a}.$$

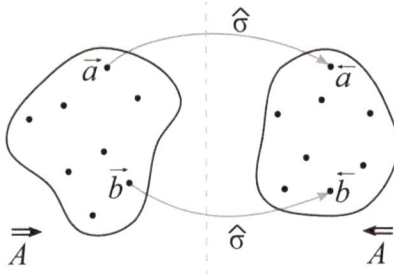

Figure 5.5 Conjugate vectors like $\overleftarrow{a}$ form a conjugate space which we denote $\overleftarrow{A}$. It is *conjugate* or *dual* to the "usual" vectors space $\overrightarrow{A}$.

We then showed that all such linear operators $\overleftarrow{a}$ behave like vectors. They can be multiplied by a number, added, written in terms of some basis, and have components that transform according to a special rule:

$$\overleftarrow{a} = a_i \overleftarrow{e}_i = a'_j \overleftarrow{e}'_j \quad \text{where} \quad \boxed{a_i = a'_j E_{ji}.}$$

We discovered a natural connection (*duality*) between every arrow-vector and its conjugate vector:

$$\overrightarrow{a} \xleftrightarrow{\hat{\sigma}} \overleftarrow{a}.$$

We can start with unary linear operators, without specifying their origin, and show that they behave like vectors. We will do it now.

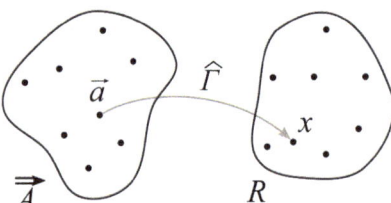

Figure 5.6 Some linear operators map vectors into numbers: $\widehat{\Gamma}\overrightarrow{a} = x$. Such operators form a vector space of their own.

Imagine *all possible* unary linear operators that map vectors into numbers (see Figure 5.6). Let us denote some such operator as $\widehat{\Gamma}$:

$$\widehat{\Gamma}\overrightarrow{a} = x_a,$$

$$\widehat{\Gamma}(\alpha\overrightarrow{a}) = \alpha(\widehat{\Gamma}\overrightarrow{a}) = \alpha x_a.$$

For $\vec{c} = \vec{a} + \vec{b}$:

$$\widehat{\Gamma}\,\vec{c} = (\widehat{\Gamma}\,\vec{a}) + (\widehat{\Gamma}\,\vec{b}) = x_a + x_b = x_c.$$

Here x_a, x_b, and x_c are real numbers.

We can demonstrate that operators like $\widehat{\Gamma}$ can be added:

$$\widehat{\Gamma} = \widehat{\Gamma}_1 + \widehat{\Gamma}_2$$

and they can be multiplied by numbers:

$$\widehat{\Gamma}_3 = \alpha\widehat{\Gamma}_1.$$

We can also find basis and expand an arbitrary operator in that basis:

$$\widehat{\Gamma} = \gamma_1\widehat{\Gamma}_1 + \gamma_2\widehat{\Gamma}_1 + \ldots + \gamma_n\widehat{\Gamma}_n = \gamma_i\widehat{\Gamma}_i$$

and establish the transformation rule for the components γ_i between different bases. In effect, we will demonstrate that unary linear operators of the same type as $\widehat{\Gamma}$ (mapping arrow-vectors into numbers) behave like vectors and therefore must be considered as such.

Describing an Operator

Remember that when we say that an operator $\widehat{\Gamma}$ is given or known, we mean that we know how it acts on *any vector* $\vec{a}$:

$$\widehat{\Gamma}\,\vec{a} = x_a.$$

Addition

If we know two operators $\widehat{\Gamma}_1$ and $\widehat{\Gamma}_2$, then making sense of their *operator sum* $\widehat{\Gamma} = \widehat{\Gamma}_1 + \widehat{\Gamma}_2$ is easy:

$$\widehat{\Gamma}\,\vec{a} = (\widehat{\Gamma}_1\,\vec{a}) + (\widehat{\Gamma}_2\,\vec{a}) = x_a + y_a,$$

where $x_a = \widehat{\Gamma}_1\,\vec{a}$ and $y_a = \widehat{\Gamma}_2\,\vec{a}$.

Multiplication

If we know the operator $\widehat{\Gamma}_1$ then making sense of the product of that operator with any number $\widehat{\Gamma} = \alpha\widehat{\Gamma}_1$ is easy:

$$\widehat{\Gamma}\,\vec{a} = \alpha(\widehat{\Gamma}_1\,\vec{a}) = \alpha x_a,$$

where $x_a = \widehat{\Gamma}_1\,\vec{a}$.

Finding Basis

Using the linearity of the operator $\widehat{\Gamma}$, we can apply it to an arbitrary vector $\vec{a}$ as follows:

$$\widehat{\Gamma}\,\vec{a} = \widehat{\Gamma}\,(a_i\vec{e}_i) = a_i\,(\widehat{\Gamma}\vec{e}_i).$$

The last expression states that to know the action of the operator $\widehat{\Gamma}$ on an arbitrary vector $\vec{a}$ it is sufficient to specify its action on all basis vectors $\vec{e}_i$. In other words, the operator $\widehat{\Gamma}$ is fully specified if we know a set of numbers

$$\boxed{\gamma_i = \widehat{\Gamma}\,\vec{e}_i.}$$

Notice how similar this last expression is to the definition of components L_{ij} of a linear operator $\widehat{L}$.

Operators and Vectors

A vector $\vec{a}$ is completely determined if we specify its components in a given basis:

$$\vec{a} = a_i\vec{e}_i.$$

A linear operator $\widehat{\Gamma}$ that maps vectors into numbers

$$\vec{a} \xrightarrow{\ \widehat{\Gamma}\ } x_a$$

is completely determined if we specify its action on basis vectors:

$$\gamma_i = \widehat{\Gamma}\,\vec{e}_i.$$

This makes the similarity between vectors and operators $\widehat{\Gamma}$ stronger.

Let's expand the input vector $\vec{a}$ in a different basis:

$$\vec{a} = a'_j\vec{e}\,'_j.$$

In this case, the action of the linear operator $\widehat{\Gamma}$ will be

$$\widehat{\Gamma}\,(a'_j\vec{e}\,'_j) = a'_j\,(\widehat{\Gamma}\vec{e}\,'_j) = a'_j\gamma'_j.$$

Here $\gamma'_j = \widehat{\Gamma}\,\vec{e}\,'_j.$

The relation between the components a_i and a_i' of the contravariant vector $\vec{a}$ is known; the relation between the values γ_i and γ_j' is then easily found:

$$\gamma_i' = \widehat{\Gamma}\vec{e}_i' = \widehat{\Gamma}(E_{ij}\vec{e}_j) = E_{ij}\gamma_j,$$

and, in a similar way:

$$\gamma_i = \widehat{\Gamma}\vec{e}_i = \widehat{\Gamma}(E_{ij}'\vec{e}_j') = E_{ij}'\gamma_j'.$$

These relations correspond to the *covariant vector.*

Covariant Vectors

The components of contravariant vectors allow us to "assemble" them from the "building blocks" – basis vectors:

$$\vec{a} = a_i\vec{e}_i.$$

The basis vectors and all other vectors constructed from them all *live in the same vector space*, which we denoted $\vec{A}$ – the space of *contravariant vectors*.

Unary linear operators, including all conjugate vectors, belong to a different vector space – *conjugate* or *dual* to $\vec{A}$. We denoted this vectors space as $\overleftarrow{A}$. Figure 5.5 illustrates this point.
This implies that it is *incorrect* to write

$$\widehat{\Gamma} = \gamma_i\vec{e}_i. \quad (incorrect!)$$

Conjugate space $\overleftarrow{A}$ has its own basis (or bases).

It is important to understand that the coefficients γ_i and γ_j' refer to bases used for contravariant vectors (bases $\{\vec{e}_i\}$ and $\{\vec{e}_i'\}$). They can also refer to bases used for covariant vectors. Let's find some such basis related to the coefficients γ_i.

Every covariant vector $\widehat{\Gamma}$ is completely determined if we know its action on *all* basis contravariant vectors $\vec{e}_i$. Therefore, to specify some basis vectors for $\widehat{\Gamma}$ we should find certain covariant vectors that can be used as "building blocks" for $\widehat{\Gamma}$. As a matter of fact, we already encountered basis covariant vectors when we discussed dol-operator and conjugate vectors of $\vec{e}_i$. We now will define similar vectors without referring to dol-operator $\widehat{\sigma}$. Specifically, the first basis vector $\widehat{\Gamma}_1$ for covariant vectors acts on $\vec{e}_i$ as follows:

$$\widehat{\Gamma}_1 \vec{e}_1 = 1 \tag{5.1}$$
$$\widehat{\Gamma}_1 \vec{e}_2 = 0 \tag{5.2}$$
$$\widehat{\Gamma}_1 \vec{e}_3 = 0 \tag{5.3}$$
$$\dots \tag{5.4}$$

Thus, $\widehat{\Gamma}_1$ returns zero for all basis vectors except for $\vec{e}_1$. Similarly, we define the second covariant basis vector $\widehat{\Gamma}_2$ to return zero for all basis vectors except for $\vec{e}_2$, and so on for other basis covariant vectors. A compact way to express this idea uses index notation:

$$\boxed{\widehat{\Gamma}_i \vec{e}_j = \delta_{ij}.}$$

Here δ_{ij} is the Kronecker delta, introduced in Chapter 2 on page 21.

Having defined this covariant basis, we can write now

$$\widehat{\Gamma} = \gamma_1 \widehat{\Gamma}_1 + \gamma_2 \widehat{\Gamma}_1 + \dots \gamma_n \widehat{\Gamma}_n = \gamma_j \widehat{\Gamma}_j.$$

Clearly,

$$\widehat{\Gamma} \vec{e}_i = \gamma_i.$$

(To demonstrate this quickly, recall that $\widehat{\Gamma} \vec{e}_i = \gamma_j \widehat{\Gamma}_j \vec{e}_i = \gamma_j \delta_{ji} = \gamma_i$.) In other words, the coefficients γ_i are also *components* of the covariant vector $\widehat{\Gamma}$ in *covariant basis* $\{\widehat{\Gamma}_i\}$.

Geometric Representation

Arrows provide a simple geometric representation of *contravariant* vectors. Now that we encountered *covariant* vectors, it is natural to ask what geometric representation covariant vectors have?

Unary linear operators map vectors into numbers. In a certain sense, they "complete" vectors to a mathematical construction that can be unambiguously assigned a number. One such construction is the oriented area (for a plane) or a volume (for three and higher dimensions).

Let's use arrows in three-dimensional space as an example. What completes a given arrow-vector $\vec{a}$ to a volume? We can build a solid figure with a well-defined volume using the arrow as the side of a cylinder, and some two-dimensional area-element as its "complement." This area-element will correspond to a linear operator that maps the vector $\vec{a}$ into a number – the volume of the solid object built using the vector and the area-element, as shown in Figure 5.7(a).

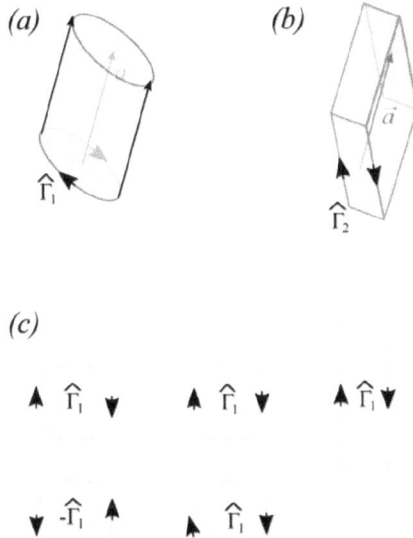

Figure 5.7 (a) A covariant vector $\widehat{\Gamma}_1$ can be represented by an oriented piece of plane with certain area. Its action on a contravariant vector $\vec{a}$ result in a number – volume of a skewed cylinder built by moving the area along the vector. (b) A covariant vector $\widehat{\Gamma}_2$ has different orientation and magnitude from $\widehat{\Gamma}_1$. (c) The shape of the conjugate vector (an oriented piece of plane) does not matter, as long as its orientation and area stay the same.

Covariant Vectors

Contravariant arrow-vectors $\vec{a}$ are oriented line segments, regardless of whether they are in a plane, in three-dimensional space, or in spaces of higher dimensions.

Covariant vectors have different "structure" for spaces of different dimensions. In three dimensions, as we saw, they are oriented areas. In a plane, they will be oriented line segments similar to arrow-vectors. In four dimensional space covariant vectors will be oriented three-dimensional volumes. Thus, unlike contravariant arrow-vectors, covariant vectors are more difficult to visualize.

The question of the geometric meaning of the addition of two covariant vectors – and other operations on them – although interesting and useful, is outside of the scope of this book.

5.7 PROJECTORS

In many problems, it is useful to take a vector $\vec{b}$ and find that part of it which will be parallel to another vector $\vec{a}$, as illustrated in the Figure 5.8. This procedure can be described using the concept of a binary operator. This operator, when given a pair of vectors $\vec{a}$ and $\vec{b}$, returns the "component" of $\vec{b}$ oriented along $\vec{a}$ – a *projection* of $\vec{b}$ onto $\vec{a}$:

$$\widehat{P}\,\vec{a}\,\vec{b} = \vec{b}_{\parallel},$$

where $\vec{a}$ is parallel to $\vec{b}_{\parallel}$.

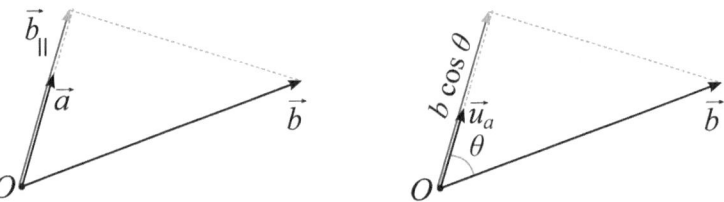

Figure 5.8 A vector $\vec{b}$ has the "component" $\vec{b}_{\parallel}$ along a given vector $\vec{a}$. The length of this component is $b\cos\theta$ where θ is the angle between vectors $\vec{a}$ and $\vec{b}$.

From the Figure 5.8, it is clear that

$$\vec{b}_{\parallel} = \vec{u}_a b\cos\theta = \frac{\vec{a}}{a}b\cos\theta,$$

where $\vec{u}_a$ is a unit-length vector parallel to $\vec{a}$.

The right-hand side of the last expression can be written using dol-operator $\widehat{\sigma}$, or even better, using the conjugate vector notation:

$$\vec{b}_{\parallel} = \frac{\vec{a}}{a^2}ab\cos\theta = \frac{\vec{a}}{a^2}(\overleftarrow{a}\,\vec{b}).$$

For each vector $\vec{a}$ there exists a corresponding operator – *projection operator* or *projector*– that projects all other vectors onto the direction specified by $\vec{a}$.

Projector Notation

A projector operator that projects any vector $\vec{b}$ onto the direction specified by a vector $\vec{a}$ will be denoted as $\underline{\underline{\widehat{A}}}$:

$$\underline{\underline{\widehat{A}}}\,\vec{b} = \frac{\vec{a}}{a^2}\,(\overleftarrow{a}\vec{b}).$$

Such projector exists for any non-zero vector $\vec{a}$:

$$\vec{a} \longrightarrow \underline{\underline{\widehat{A}}}.$$

In this notation, the same (but capitalized) letter is used for the projector as for the vector. In addition, the capitalized letter is doubly underlined to remind that we project onto the direction parallel to the specified vector.
Similarly, we will have other projectors

$$\vec{b} \longrightarrow \underline{\underline{\widehat{B}}}, \quad \vec{c} \longrightarrow \underline{\underline{\widehat{C}}}, \ldots$$

Projector $\underline{\underline{\widehat{A}}}$ corresponding to the vector $\vec{a}$ acts on an arbitrary vector $\vec{b}$ as follows:

$$\underline{\underline{\widehat{A}}}\,\vec{b} = \frac{\vec{a}}{a^2}\,(\overleftarrow{a}\vec{b}).$$

In the argument-free notation (see note on page 26), this operator takes the following form:

$$\underline{\underline{\widehat{A}}} = \frac{\vec{a}}{a^2}\,(\overleftarrow{a}) = \frac{\vec{a}\,\overleftarrow{a}}{a^2}.$$

Note: The order of the vectors $\vec{a}$ and $\overleftarrow{a}$ in the last expression is very important because it has a completely different meaning from the expression $\overleftarrow{a}\vec{a}$. Indeed, as we agreed, the expression

$$\overleftarrow{a}\vec{a} = \vec{a}\cdot\vec{a} = a^2$$

yields the length squared of the vector $\vec{a}$. In contrast, the expression

$$\boxed{\vec{a}\,\overleftarrow{a}}$$

works as an operator!

We obtained interesting and useful expression for an operator that accepts a vector $\vec{b}$ and produces another vector as the result. This expression involves some kind of "multiplication" of a vector $\vec{a}$ and its conjugate $\overleftarrow{a}$:

$$\vec{a}\,\overleftarrow{a}.$$

It is our first encounter with *tensor product*. We will learn more about this new type of multiplication below.

5.7.1 Projector Components

To find the components of a projector operator

$$\widehat{\underline{\underline{A}}} = \frac{\vec{a}\,\overleftarrow{a}}{a^2}$$

in a given basis, we apply it to the basis vectors:

$$\widehat{\underline{\underline{A}}}\,\vec{e}_i = \frac{\vec{a}}{a^2}\,\overleftarrow{a}\,\vec{e}_i = \frac{a_i}{a^2}\,\vec{a}.$$

Expanding the vector $\vec{a}$ in the same basis, we arrive at

$$\widehat{\underline{\underline{A}}}\,\vec{e}_i = \frac{a_i a_j}{a^2}\,\vec{e}_j = \underline{\underline{A}}_{ij}\,\vec{e}_j,$$

from which follows the expression for the components

$$\boxed{\underline{\underline{A}}_{ij} = \frac{a_i a_j}{a^2}.}$$

Exercise 5.4 *Using the components of a projector*

$$\underline{\underline{A}}_{ij} = \frac{a_i a_j}{a^2},$$

calculate its determinant.

Symmetry of Projectors

From the expression for the components of a projector follows that

$$\underline{\underline{A}}_{ij} = \underline{\underline{A}}_{ji},$$

which means that to fully specify its components, we need only 3 numbers (for vectors in a plane we need $\underline{\underline{A}}_{11}$, $\underline{\underline{A}}_{12}$, and $\underline{\underline{A}}_{22}$), as opposed to 4 numbers required to specify a general linear operator.

5.7.2 Composition of Projectors*

The idea of composing two functions - discussed in subsection 2.3.4 on page 25 - can be extended to linear operators. That is, some types of linear operators can be composed. Indeed, suppose we have a linear operator

$$\widehat{L}\,\vec{a} = \vec{b}$$

and

$$\widehat{M}\,\vec{b} = \vec{c}.$$

We can apply the operators $\widehat{L}$ and $\widehat{M}$ sequentially:

$$\widehat{M}\,(\widehat{L}\,\vec{a}) = \vec{c}.$$

This way we obtained a new operator $\widehat{K}$ which we call *composition of linear operators* $\widehat{L}$ and $\widehat{M}$. The same notation for composition of operators as for the composition of functions can be used:

$$\widehat{K} = \widehat{M} \circ \widehat{L}.$$

Now we can find the components of the operator $\widehat{K}$. On the one hand,

$$\widehat{K}\,\vec{e}_i = K_{ij}\vec{e}_j.$$

On the other,

$$\widehat{K}\,\vec{e}_i = \widehat{M}\,(\widehat{L}\,\vec{e}_i) = \widehat{M}\,(L_{iq}\,\vec{e}_q).$$

Using the linearity of the operator $\widehat{M}$, we can write

$$\widehat{K}\,\vec{e}_i = L_{iq}\,(\widehat{M}\,\vec{e}_q) = L_{iq}(M_{qj}\,\vec{e}_j).$$

We arrive at the following expression of the components of $\widehat{K}$:

$$K_{ij} = L_{iq}M_{qj},$$

where the summation over q is implied, according to Einstein's summation rule. Now let us apply this result to projectors.

Suppose we want to project a vector $\vec{c}$ first on the vector $\vec{a}$, and then project the result onto the vector $\vec{b}$. We can do it by sequential application of two projectors:

$$\underline{\underline{\widehat{A}}} = \frac{\vec{a}\,\overleftarrow{a}}{a^2} \quad \text{and} \quad \underline{\underline{\widehat{B}}} = \frac{\vec{b}\,\overleftarrow{b}}{b^2}$$

to the vector $\vec{c}$:

$$\widehat{\underline{\underline{B}}}(\widehat{\underline{\underline{A}}}\,\vec{c}) = (\widehat{\underline{\underline{B}}} \circ \widehat{\underline{\underline{A}}})\,\vec{c}.$$

Composing two linear operators we obtain another linear operator:

$$\widehat{L} = \widehat{\underline{\underline{B}}} \circ \widehat{\underline{\underline{A}}}.$$

The components of the product operator $\widehat{L}$ can be expressed in terms of the components of the factors – projectors $\widehat{\underline{\underline{A}}}$ and $\widehat{\underline{\underline{B}}}$:

$$L_{ij} = \widehat{\underline{\underline{A}}}_{ik}\,\widehat{\underline{\underline{B}}}_{kj} = \frac{a_k b_k}{a^2 b^2} a_i b_j.$$

The expression $a_k b_k$ is recognized as the scalar product of the vectors $\vec{a}$ and $\vec{b}$ in orthonormal basis, so the components L_{ij} are simply given by

$$L_{ij} = \lambda\, a_i b_j, \qquad \lambda = \frac{\vec{a} \cdot \vec{b}}{a^2 b^2},$$

where λ is a scalar value – a number.

Note: The operator $\widehat{L} = \widehat{\underline{\underline{B}}} \circ \widehat{\underline{\underline{A}}}$ is no longer a projector in the sense in which it was defined earlier. There is no vector $\vec{d}$ such that

$$L_{ij} = \frac{d_i d_j}{d^2}.$$

The following exercise explores this point.

Exercise 5.5 *(a) Show that a projector*

$$\widehat{\underline{\underline{A}}} = \frac{\vec{a}\,\overleftarrow{a}}{a^2}$$

has the following property:

$$\widehat{\underline{\underline{A}}} \circ \widehat{\underline{\underline{A}}} = \widehat{\underline{\underline{A}}}.$$

(b) Does the result of composition $\widehat{L} = \widehat{\underline{\underline{B}}} \circ \widehat{\underline{\underline{A}}}$ have this property?

Exercise 5.6 *Consider the composition*

$$\widehat{M} = \widehat{\underline{\underline{A}}} \circ \widehat{\underline{\underline{B}}}.$$

Find its components and compare them to the components of

$$\widehat{L} = \widehat{\underline{\underline{B}}} \circ \widehat{\underline{\underline{A}}}.$$

5.8   TENSOR PRODUCT

We arrived at an extremely important idea that allows "building" tensors of various kinds from simple "ingredients," such as vectors. To begin, let us take a closer look at a projector operator:

$$\widehat{\underline{\underline{A}}} = \frac{1}{a^2}\vec{a}\overleftarrow{a}, \quad \underline{\underline{A}}_{ij} = \frac{1}{a^2}a_i a_j.$$

From the first expression, written without the components, it is clear that the operator $\widehat{\underline{\underline{A}}}$ involves a contravariant vector $\vec{a}$ and its covariant conjugate $\overleftarrow{a}$. This distinction is absent in the second expression for the components of the operator $\underline{\underline{A}}_{ij}$, making the expression for the components misleading. To fix this, a special notation for components is introduced. In this notation, the components of contravariant vectors are written using superscripts:

$$\vec{a} \quad \longrightarrow \quad a^i,$$

while the components of covariant vectors are written in the "usual" way, as subscripts:

$$\overleftarrow{a} \quad \longrightarrow \quad a_i.$$

With this in mind, the components of the projector $\widehat{\underline{\underline{A}}}$ are written in the following way:

$$\boxed{\underline{\underline{A}}^{i\,\circ}_{\;\circ\, j} = \frac{1}{a^2}a^i a_j.}$$

Now it should be clear that in the last expression vectors of different kinds are used: one contravariant, and the other covariant. The little grey circles serve visual purpose only, they help separate the first contravariant index from the second covariant one.

Clash of Notation

The use of superscripts to denote components of contravariant vectors leads to the clash of notations. For example, given an expression a^2, how should we understand it: Is it the length squared of a vector, or is it the second component of a contravariant vector? Surprisingly, this is not a serious problem

at all, since the meaning of superscripts is usually clear from the context in which such expressions appear.

5.8.1 Tensor Product 1

In the expressions

$$\widehat{\underline{A}} = \frac{1}{a^2}\vec{a}\overleftarrow{a}, \quad \underline{A}^{i\,\bullet}_{\bullet\,j} = \frac{1}{a^2}a^i a_j$$

the combination of vectors $\vec{a}\overleftarrow{a}$ and their component expression $a^i a_j$ represent a mathematical object – operator in this case – that is neither a vector, nor a number. Such "amalgamation" of two vectors into a tensor is called a *tensor product*. Tensor product is a simple and versatile way to construct tensors.

Special notation for the tensor product of two vectors exists:

$$\boxed{\vec{a}\overleftarrow{a} = \vec{a} \otimes \overleftarrow{a}.}$$

This separate notation may appear redundant, since we understand from the order of the vectors $\vec{a}$ and $\overleftarrow{a}$ that the expression on the left is *not a scalar product*. However, using the infix operator $\otimes$ is convenient because it allows writing other types of tensor products with ease and consistency.

5.8.2 Tensor Product 2

Using the tensor product notation, we can write

$$\overleftarrow{a} \otimes \vec{a} \quad \text{or, more generally,} \quad \overleftarrow{b} \otimes \vec{a}.$$

These expressions *represent tensors* and not scalar products $\overleftarrow{b}\vec{a} = \vec{b} \cdot \vec{a}$. The components of a tensor $T = \overleftarrow{b} \otimes \vec{a}$ are as follows:

$$T_{i\,\bullet}^{\bullet\,j} = b_i a^j.$$

The important fact is reflected in the position of indices of the tensor T: It behaves like a covariant vector in the first index, and as a contravariant vector in the second index.

5.8.3 Tensor Product 3

With the help of the infix operator $\otimes$ we can write, without creating ambiguity, a tensor product of two contravariant vectors:

$$T = \vec{a} \otimes \vec{b}, \quad T^{ij} = a^i b^j.$$

The positions of the indices reflect the fact that this kind of tensor is contravariant in both of them.

A simplified notation is sometimes used:

$$\vec{a} \otimes \vec{b} = \vec{a}\,\vec{b},$$

but it may lead to confusion, since the expression $\vec{a}\,\vec{b}$ is too similar to the scalar product $\vec{a} \cdot \vec{b}$, especially when using handwriting. We will avoid this simplified notation.

5.8.4 Tensor Product 4

The last kind of tensor that we can construct from vectors is given by the tensor product of two covariant vectors:

$$T = \overleftarrow{a} \otimes \overleftarrow{b}, \quad T_{ij} = a_i b_j.$$

Again, one may encounter expressions like $\overleftarrow{a}\overleftarrow{b}$, but we will prefer to use the infix operator $\otimes$.

Exercise 5.7 *Write the transformation rules for tensors of all four kinds considered above.*

5.9 TENSORS DEFINED

We are now in a good position to summarize our understanding of tensors. Before we do this, let's quickly review the path we took to reach this position.

Having defined contravariant and covariant vectors, we examined the natural idea of *operators* – functions on vectors. We focused on an important class of operators called *linear operators*.

We studied linear operators that map vectors to other vectors, like rotation, and, having examined the transformation of operator components, derived the first type of transformation (see Section 4.7). This type of transformation corresponds to the tensor of a mixed kind: contravariant in the first index and covariant in the second index. Later we encountered many operators of this kind – projector operators (see Section 5.7.)

Projector operators are unary linear operators. They are built upon the bilinear dol-operator $\widehat{\sigma}$. This binary operator introduced to us the idea of conjugate space and unary linear operators mapping vectors into numbers.

From further analysis of projector operators, we arrived at the idea of tensor product and unlocked a key method of building tensors of various kinds. Using a pair of vectors, we listed four different kinds of tensors: covariant-covariant, covariant-contravariant, contravariant-covariant, and contravariant-contravariant. All these tensors can be viewed as operators acting on vectors, either covariant or contravariant, depending on the tensor type.

Having reviewed our steps, we can define tensors as follows:

Tensors

Tensors are *mathematical objects* with the following essential properties:

- Tensors can be combined (added) pairwise to yield another tensor.

- Tensors can be multiplied by real numbers to yield another tensor.

- Tensors can be represented via components, written relative to some basis.

- When the basis changes, components of tensors transform in a very specific way, to ensure that *tensors remains the same.*

This definition is deliberately analogous to the definition of vectors. In some sense tensors are *next tier vectors*. Tensors are mathematical objects following vectors in the ladder of abstraction and power, like vectors are mathematical objects following numbers in the ladder of abstraction and power.

5.9.1 Other Definitions

Let us revisit the definitions of tensors given in the introduction. The first one read:

Tensors Definition 1

Tensor on a vector space V over a field k is an element t of the vector space

$$T^{p,q}(V) = (\otimes^p V) \otimes (\otimes^q V^*),$$

where $V^* = \mathrm{Hom}(V, k)$ is the dual space of V.

In this definition the set of vectors V, which we denoted as $\overset{\Rightarrow}{A}$, is called *vector space*. Recall that vectors and operation with vectors require numbers. Numbers, which can be added, multiplied in a usual way, form what mathematicians call *field*. Therefore, all vectors taken together are technically called "vector space V over a field k."

Next, the use of the operation of tensor product $\otimes$ in the definition makes more sense, since it is the basic way to build tensors from vectors. Notice that two types of vectors are mentioned: one from the vector space V, and the other from its conjugate (dual) – vector space V^*; in our notation $V^* = \overset{\Leftarrow}{A}$. The conjugate vectors behave very much like vectors from the original vector space: they can be added, multiplied by numbers, expanded in their own basis, and so on. This close correspondence between vector space and its conjugate is called *homomorphism*[3] and denoted $\mathrm{Hom}(V, k)$.

Tensor of the kind $T^{p,q}(V) = (\otimes^p V) \otimes (\otimes^q V^*)$ has p contravariant indices and q covariant indices. We mostly dealt with tensors of the following types: $T^{1,1}$, $T^{0,2}$, $T^{2,0}$.

Finally, tensors are vectors because they can be added, multiplied by numbers, have components, and have vector-like transformation rules. Tensors of a given type, e.g. $T^{2,1}$, taken together, form a vector space; there is a separate vector space for each type of tensor.

The second definition stated:

Tensors Definition 2

An nth-rank tensor in m-dimensional space is a mathematical object that has n indices and m^n components and obeys certain transformation rules.

Here tensor is mentioned in connection with some m-dimensional space, which we recognize as the underlying vector space $\overset{\Rightarrow}{A}$. In our case, the number of dimensions, given by the number of independent directions in a plane, equals 2.

---

[3]From Greek *homos* (same) and *morphe* (shape).

The rank of a tensor is given by the number of indices or the number of vectors that go into the tensor product. For example:

$$L = \overleftarrow{a} \otimes \overleftarrow{b} \longrightarrow L_{ij} \quad \text{is the tensor of the second rank,}$$

$$M = \overleftarrow{a} \otimes \overleftarrow{b} \otimes \overrightarrow{c} \longrightarrow M_{i\,j\,\bullet}^{\bullet\,\bullet\,k} \quad \text{is the tensor of the third rank.}$$

Since the value of each index runs from 1 to m, where m is the dimension of the vector space, tensors of the rank n should have m^n components in total. The essential property of tensor components is how they transform when the basis changes – *the tensor transformation rule*.

The last definition, given in the introduction, looks as follows:

Tensors Definition 3

Just as a *vector* is a mathematical quantity that describes translations in two- or three-dimensional space, a tensor is a mathematical quantity used to describe general transformations in n-dimensional space. Precisely, if the locations of points in n-dimensional space are given in one coordinate system by $(x^1, x^2, \ldots, x^n)$ and in a transformed coordinate system by $(y^1, y^2, \ldots, y^n)$ (it is convenient to use superscripts rather than subscripts), then a "rank 1 contravariant tensor" is a quantity T, with single components, that transforms according to the rule:

$$T_{new}^i = \sum_{r=1}^{n} \frac{\partial y^i}{\partial x^r} T^r.$$

We now know that tensors can be used to describe the transformation of one vector into another (e.g., projectors). Similar ideas can be applied to vectors in spaces with higher dimensions. In this sense, tensors can describe general transformations in n-dimensional space.

The essential property of tensor components is their transformation rule. In this definition a transformation rule for a tensor of the first rank (a "usual" vector) is given:

$$T'^i = Z^i_{\ r} T^r,$$

where the set of numbers $Z^i_{\ r}$ describes the relation between the "old" basis (coordinates x) and the "new" basis (coordinates y).

CHAPTER HIGHLIGHTS

- *Two vectors can be compared for similarity by calculating the "degree of overlap." The longer two vectors are and the closer their mutual direction – the greater the overlap is.*

- *Degree of overlap can be described by a bilinear operator $\widehat{\sigma}$. This operator is closely related to the concept of the scalar product of two vectors.*

- *When scalar product (or, equivalently, degree of overlap) is defined for vectors, each vector receives a "special relative" – conjugate vector – that lives in different vector space, called conjugate or dual space.*

- *When the degree-of-overlap operator $\widehat{\sigma}$ is partially applied, the result is a unary linear operator that yields a number for every input vector. Importantly, such an operator is also a vector, albeit not an arrow-like vector.*

- *Unary linear operators that act on arrow-like vectors behave like vectors themselves and form a vector space of their own. The latter is conjugate to the "usual" space of arrow-like vectors.*

- *If arrow-like vectors are contravariant vectors, then their conjugate counterparts are covariant vectors. Covariant vectors can be represented geometrically as oriented area elements (for three dimensional space).*

- *Projectors are unary linear operators that act on input arrows to yield another arrow that is parallel to a certain direction. Projectors are degenerate operators.*

- *Projectors can be written using the efficient tensor product notation.*

- *Tensor product is a simple and powerful way to build up tensors of any rank and any kind from a number of covariant and contravariant vectors.*

- *Tensors are mathematical objects with many properties similar to vectors. However, their rank is higher and tensors can be used to express linear relations between vectors.*

CHAPTER 6

Applications of Tensors

W E ARE NOW READY TO APPRECIATE HOW tensors are used in "real life." In this chapter we will encounter examples of tensors that are used in mathematics, physics, and engineering. Before we get to the examples of tensors, one more helpful notation must be explained.

δ-Notation

When a quantity x changes by a tiny amount, we will denote the change using the small Greek letter δ (delta) as follows:

$$\boxed{\delta x \text{ - tiny change of } x.}$$

For example, for the earth going around the sun in 365 days, one second elapsed on a clock can be considered a tiny change δt. When a drop of water falls into a nearly full bucket the mass of the latter changes by a tiny amount δm, and so on.

That's all there is to δ-notation. We are not going into the realm of calculus, where mathematicians talk about infinitesimal quantities and limits; we will be simply using "tiny changes." Now on to tensors.

6.1 FAMOUS TENSORS

We will study several examples of tensors that readers most likely encounter in geometry, physics, and engineering. The material in the previous chapters should be enough to prepare readers to deal with tensors of any kind. However, we will limit considerations to simple tensors of lower ranks.

DOI: 10.1201/9781003620365-6

6.1.1 Metric Tensor

Metric tensor is used to determine distances between pairs of points in space. A distance between two points is equal to the length of a vector connecting them, as shown in Figure 6.1. For a vector

$$\vec{d} = \vec{b} - \vec{a}, \quad d^i = b^i - a^i$$

its length squared is given by the scalar product

$$d^2 = \vec{d} \cdot \vec{d} = \hat{\sigma}\, \vec{d}\, \vec{d}.$$

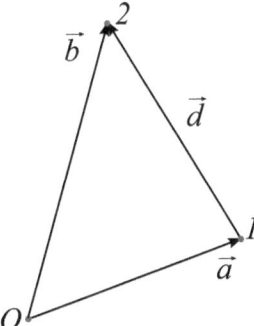

Figure 6.1 Distance between two points (1 and 2) equals to the lengths of the vector $\vec{d}$ connecting these points.

Using components in an arbitrary basis (not orthonormal), the length squared is written as

$$d^2 = \hat{\sigma}\, (d^i \vec{e}_i)\, (d^j \vec{e}_j) = d^i d^j\, (\hat{\sigma}\, \vec{e}_i\, \vec{e}_j).$$

The set of values

$$\eta_{ij} = \hat{\sigma}\, \vec{e}_i\, \vec{e}_j = \vec{e}_i \cdot \vec{e}_j$$

corresponds to the components of a special tensor – *metric tensor*. The transformation rule of these components is easily found by expanding the "old" basis vectors in terms of the "new" (primed) basis, and using the linearity of $\hat{\sigma}$:

$$\eta'_{ij} = \hat{\sigma}\, \vec{e}\,'_i\, \vec{e}\,'_j = \hat{\sigma}\, (E_{im} \vec{e}_m)\, (E_{jn} \vec{e}_n) = E_{im} E_{jn} \eta_{mn}.$$

This is the transformation rule of a covariant-covariant tensor of the second rank. The metric tensor has to be of this kind since it maps a contravariant-contravariant tensor

$$\vec{d} \otimes \vec{d}, \quad d^i d^j$$

into a scalar. Each index of the metric tensor must transform in a way that "compensates" the contravariant transformation of $\vec{d}$ in the tensor product $\vec{d} \otimes \vec{d}$. This is analogous to how a covariant vector $\overleftarrow{b}$ maps a contravariant vector $\vec{a}$ into a number:

$$\overleftarrow{b}\,\vec{a} \longrightarrow x$$

$$\eta\left(\vec{a} \otimes \vec{a}\right) \longrightarrow y.$$

Exercise 6.1 *Although the primary use of the metric tensor is to calculate distances between a pair of points connected by a vector $\vec{d}$, it can be applied to any pair of contravariant vectors:*

$$\eta\left(\vec{a} \otimes \vec{b}\right).$$

a) What is the meaning of this operation? b) What are the components of the metric tensor in an orthonormal basis?

In most elementary problems of geometry and physics, a coordinate system in a plane is Cartesian and basis vectors are orthonormal. As the result, the components of the metric tensor are trivial – zeros and ones – and are the same everywhere in the plane.

When non-Cartesian coordinates are used in a plane, e.g. polar coordinates shown in Figure 6.2, the basis vectors are aligned with the coordinate grid[1] and have different orientations in different points. In this case the components of the metric tensor η_{ij} change from point to point to ensure that the lengths of the vector $\vec{d}$

$$d^2 = \eta_{ij} d^i d^j$$

remains the same.

---

[1]In principle, it is possible to have basis vectors "decoupled" from the coordinate system, but this is not very convenient.

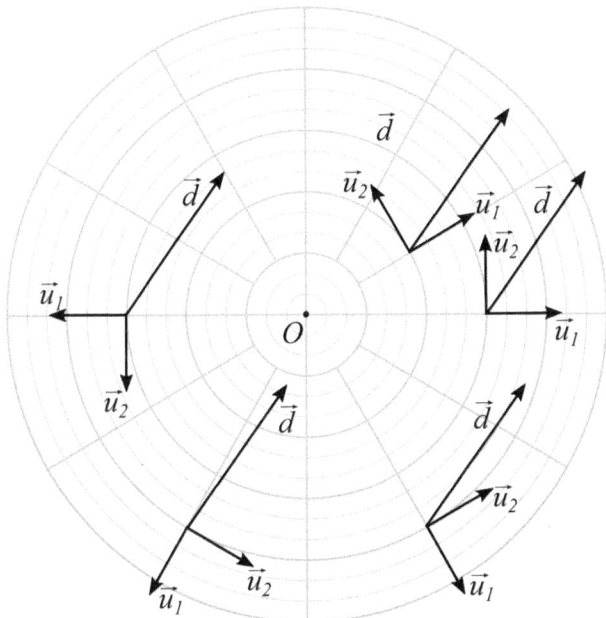

Figure 6.2 Components of the same vector $\vec{d}$ differ in various points of the plane because the unit basis vectors $\{\vec{u}_1, \vec{u}_2\}$ in those points are aligned with the grid of the polar coordinate system. The vector $\vec{u}_1$ points in the direction of increasing distance r; the vector $\vec{u}_2$ points in the direction of increasing angle θ.

Moreover, for surfaces more sophisticated than a plane (e.g., sphere, paraboloid, saddle-like surface, and myriad of others), it is impossible to use coordinate system and basis such that the metric tensor is constant. The components of the metric tensor will vary across the surface to reflect the real, and not a merely "coordinate induced," difference of a surface from a plane. In other words, variations of the components of the metric tensor indicate that the surface is *curved.*

As an example, consider the two-dimensional surface of a sphere, shown in Figure 6.3. Each point on the surface can be located using a pair of coordinates – the angle $\theta = x^1$ complimentary to the latitude, and the longitude angle $\phi = x^2$. If a pair of close points on the sphere have

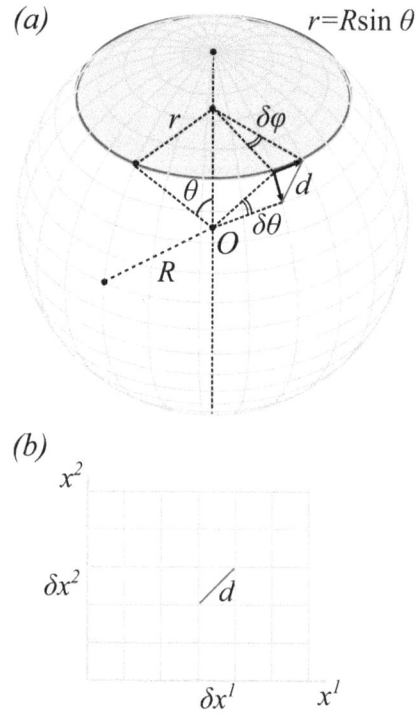

(a)

$r = R\sin\theta$

(b)

x^2

δx^2

d

δx^1 x^1

Figure 6.3 (a) Positions of points on the surface of a sphere can be specified using two coordinates: angle $\theta = x^1$ and $\phi = x^2$. (b) A part of a sphere covered by a coordinate grid (x^1, x^2) can be represented using perpendicular axes, similar to Cartesian coordinates (x, y). Distance between two points on a uniformly spaced coordinate grid differs from point to point: $d^2 = R^2(\delta x^1)^2 + R^2\sin^2 x^1 (\delta x^2)^2$.

coordinates

$$
\begin{array}{ccc}
1 & \longrightarrow & (\theta, \phi), \\
2 & \longrightarrow & (\theta + \delta\theta, \phi + \delta\phi),
\end{array}
$$

then the distance squared between these points is given by

$$d^2 = R^2(\delta\theta)^2 + R^2\sin^2\theta(\delta\phi)^2.$$

This formula is obtained by applying Pythagoras theorem to the tiny right triangle with the sides indicated using arrows in Figure 6.3. The length of the side resulting from the change of the coordinate ϕ is $r\delta\phi = R\sin\theta\delta\phi$; the length of the side resulting from the change of the coordinate θ equals $R\delta\theta$.

Using the uniform notation for coordinates, the distance squared is written as

$$d^2 = R^2(\delta x^1)^2 + R^2 \sin^2 x^1 (\delta x^2)^2.$$

Comparing this to the Cartesian expression $d^2 = (\delta x^1)^2 + (\delta x^2)^2$, we can see that not all components of the metric tensor in the spherical coordinate basis are constant. Namely, the component $\eta_{22} = R^2 \sin^2 x^1$ depends on the coordinate $x^1 = \theta$.

Einstein's Equations for Gravity

The notion of the metric tensor is central to the Einstein's General Theory of Relativity. The theory describes the effects of gravity in terms of geometrical notions – curvature of the universe at various points of space and time due to the effects of energy and motion in those points.

The main equation of the theory can be written using index notation as follows:

$$R_{ij} - g_{ij}(\Lambda - R/2) = \kappa T_{ij}.$$

Here g_{ij} is the metric tensor, R_{ij} is the curvature tensor, and T_{ij} is the tensor of energy and motion. The scalar quantity R is the numerical characteristic of the curvature tensor R_{ij}, while Λ is the so-called *cosmological term* required to describe the measured expansion of the observable universe.

The goal of the theory is to find metric tensors satisfying the equation above and then understand what geometric shapes those metric tensors describe. The difficulty comes with the fact that both the curvature tensor R_{ij} and the tensor of energy and motion T_{ij} depend on the metric tensor, making Einstein's equations highly *nonlinear*.

6.1.2 Anisotropy Tensor

Anisotropy tensor is a general term for various tensors used in physics to describe properties of materials like crystalline solids. Many physical properties – mechanical, optical, electronic, thermal – describe the response of the material to external "forces" or perturbations. Mathematical description of such responses requires tensors.

To understand the general idea, let us consider a simple situation, depicted in Figure 6.4. Suppose that a tree bends in the wind, so that when the wind blows in the direction of the x axis, the displacement of the tree-top is also along the x axis, with the magnitude proportional to

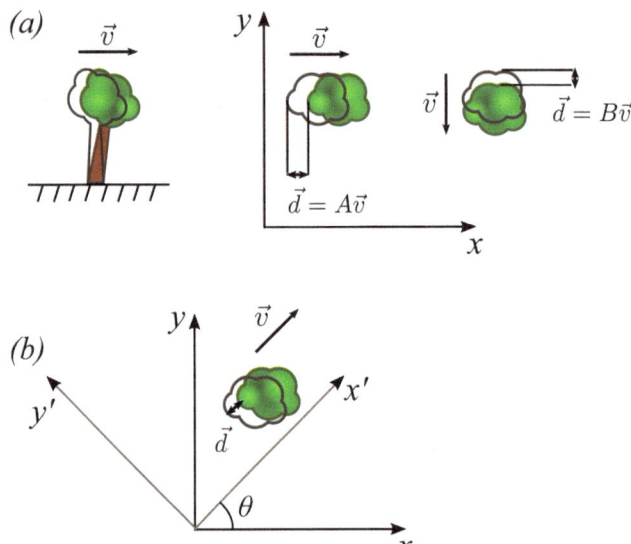

Figure 6.4 (a) A tree bends in the wind by the amount and the direction determined by the velocity vector $\vec{v}$; On the right, the top view of the tree is given. (b) For an arbitrary direction of the wind, the displacement of the tree-top $\vec{d}$ will not be parallel to the direction of the wind, but it will still be proportional to the wind velocity $\vec{v}$.

the magnitude of the wind's velocity $\vec{v}$:

$$\vec{d} = d_x \vec{u}_1 = A v_x \vec{u}_1 = A\vec{v}.$$

Next, suppose that when the wind blows along the y axis, the tree-top is also displaced in the direction of the y-axis:

$$\vec{d} = d_y \vec{u}_2 = B v_y \vec{u}_2 = B\vec{v}.$$

In both cases the displacement vector $\vec{d}$ is parallel to the vector of wind's velocity.

For a general direction of the wind, the magnitude of the tree-top displacement will be proportional to the magnitude of the wind's velocity, but *the direction of the displacement will differ from the direction of the wind*:

$$d \propto v, \quad \vec{d} \nparallel \vec{v}.$$

Indeed, for a wind vector

$$\vec{v} = v_x \vec{u}_1 + v_y \vec{u}_2 = \vec{v}_1 + \vec{v}_2,$$

the "response" of the tree-top will be different for different components of the wind vector:

$$\vec{d} = \vec{d}_1 + \vec{d}_2 = A\vec{v}_1 + B\vec{v}_2 = Av\cos\theta\vec{u}_1 + Bv\sin\theta\vec{u}_2 = d_x\vec{u}_1 + d_x\vec{u}_2.$$

Clearly,

$$\frac{d_y}{d_x} = \frac{B\sin\theta}{A\cos\theta} \neq \tan\theta.$$

Thus, although the displacement magnitude d is still proportional to the magnitude of the wind's velocity v, the direction of the displacement no longer coincides with the direction of the wind. This fact can be expressed using tensor notation:

$$d^i = T^i{}_j v^j. \tag{6.1}$$

In the special coordinates, considered at the beginning of this problem, the components of the tensor T are simple:

$$T^1{}_1 = A, T^1{}_2 = 0, T^2{}_1 = 0, T^2{}_2 = B.$$

In all other coordinate systems and bases, the components of the "response tensor" T can be found using the transformation rule for the contravariant-covariant tensor of the second rank.

Expressions similar to the equation (6.1) can be written for a variety of physical phenomena. We will consider a couple of examples next.

Mechanics: Stress Tensor

Mechanics of elastic media use many tensor tools. One of the basic tensors is the *stress tensor*.[2] This tensor describes the distribution of mechanical stress inside a deformed elastic body, as illustrated in the Figure 6.5.

An elastic ball, shown in the Figure 6.5(a), can be squeezed by external forces, resulting in the change of shape (*deformation*) and the appearance of mechanical stress inside the ball; see Figure 6.5(b). In general, the induced stress will change from point to point inside the

---

[2] Also known as *Cauchy stress tensor* or *true stress tensor*.

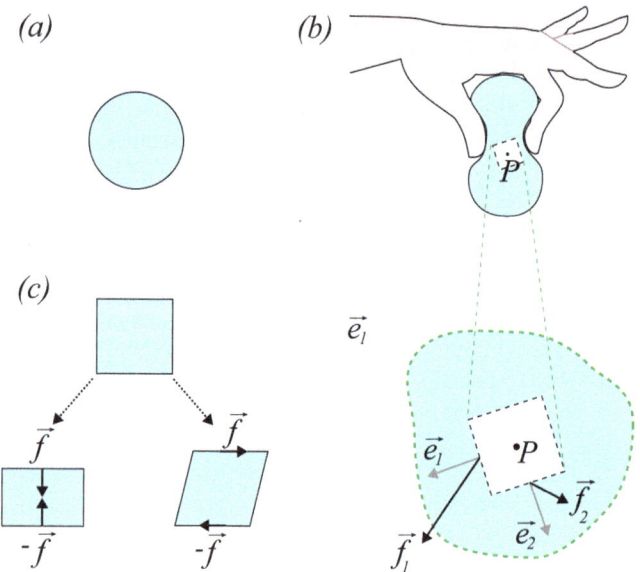

Figure 6.5 Stress tensor describes the distribution of stress inside an elastic medium. (a) An elastic ball without stress and deformation. (b) Applying external forces deforms elastic body and creates mechanical stress inside. (c) Forces perpendicular to the sides of the square lead to stretching or compression, whereas forces parallel to the surfaces result in shear. (d) Total force applied to each side of the square removed from the stressed body is given by the stress tensor: $\vec{f}_i = \widehat{\sigma}\vec{e}_i$. See text for more details.

deformed ball. To describe the stress around a given point P, we can imagine that a small part of the body is removed, leaving a tiny square-shaped hole. If nothing is done, the empty part of the ball around the point P will not remain square, due to the "forces" acting within the body and at the boundary of the hole. To keep the hole square, we must compensate the forces due to mechanical stress and apply the balancing forces $\vec{f}_1$, $\vec{f}_2$, $\vec{f}_3$, $\vec{f}_4$ to each side of the square[3]. Only two such forces are shown in the Figure 6.5(b) for simplicity.

The direction and magnitude of a force needed for a given side can be found as follows: First, find the unit-length vector $\vec{e}_i$ perpendicular

---

[3]For a three dimensional ball, the shape of the hole will be a cube, and the number of forces will be 8 – one for each face of the cube.

to the side. Second, calculate the force using the Cauchy stress tensor:

$$\vec{f}_i = \widehat{\sigma}\,\vec{e}_i.$$

The traditional notation for mechanical stress tensors is the Greek letter sigma – σ. It should not be confused with our notation for the dol-operator defined earlier (see Section 5.1).

It is easy to understand why the "balancing forces" are not, in general, pushing perpendicular to the sides of the square. The idea is illustrated in Figure 6.5(c). An elastic square (cube) can be deformed in two basic ways: 1) A square can be squeezed by forces perpendicular to the sides (*normal stress*); 2) A square can be deformed into a parallelogram by forces parallel to the sides (*shear stress*). Both types of stress can exist at the same time, resulting from forces directed at an arbitrary angle relative to the vector $\vec{e}_i$ perpendicular to the sides.

Electronics: Mobility Tensor

Many materials conduct electric current. An important characteristic of any such material is its *electrical resistance*. When a voltage V is applied to a piece of conducting material, the current I will flow between the terminals, as shown in Figure 6.6.

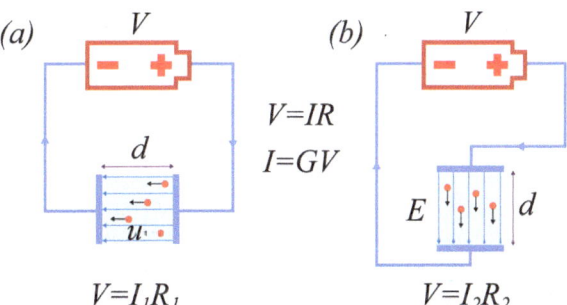

Figure 6.6 Electron mobility tensor describes the response of charge carriers inside an anisotropic material to external electric field. (a) When electric field $\vec{E}$ – and corresponding voltage V is applied to a body, the motion of electric charges (electric current I) is induced. (b) Applying the same electric field (and voltage) in different directions may result in different magnitudes of the current. This is due to the *mobility* of electric charges being different for different directions of motion.

The basic law that relates the voltage V and the current I between the terminals is *Ohm's law*:

$$V = IR.$$

Here R is the electrical resistance of a given piece of material.

For the same material, currents flowing in different directions may experience different resistances even for the same geometrical shape. In the example shown in Figure 6.6(a,b), for currents flowing horizontally and vertically through a square we can write

$$V = I_1 R_1 \text{ and } V = I_2 R_2.$$

We can take another view on the movement of electric charge through the material if we rewrite Ohm's law as follows:

$$I = GV.$$

Here instead of resistance, we use an equally useful physical parameter called *conductance* G. In a certain sense, conductance is more fundamental since it is closely related to basic physical laws that govern the motion of electric charges.

Electric current is the flow of a large number of charge carriers, such as electrons or ions, shown as red dots in Figure 6.6(a,b). The current I is proportional to the average speed u of the carriers through the material:

$$I \propto u.$$

The carriers, in their turn, move because there is an electric field E between the terminals due to applied voltage V. The average speed u of charge carriers is often simply proportional to the electric field:

$$u = \mu E,$$

where the coefficient μ is called *mobility* of charge carriers.

Now for anisotropic materials, the relation between the average velocity $\vec{u}$ of charge carriers and the applied electric field $\vec{E}$ can be written using the concept of *mobility tensor*:

$$\vec{u} = \widehat{\mu}\,\vec{E}.$$

The mobility tensor expresses how easy it is to make electrons move in a given direction by applying an external electric field $\vec{E}$.

Let us summarize: Applying voltage between terminals creates an electric field $\vec{E}$ in a given direction. The electric field is proportional to the voltage between the terminals: $E = V/d$. The electric field leads to the "mass migration" of charge carriers with the average speed

$$u = \mu E = \mu V/d.$$

This type of motion is called electric current:

$$I \propto u \quad \longrightarrow \quad I \propto \frac{\mu}{d} V.$$

From the last expression, we can see how the relationship $I = GV$ or Ohm's law $V = IR$ appears. Furthermore, because the mobility $\widehat{\mu}$ is in general a tensor, the measured resistance of a given piece of material may be different for different directions of applied voltage drop V.

Anisotropy Tensors in Physics

Besides two examples of tensors (stress and mobility) given above, there are many other tensors used in physics. Some tensors are similar to stress and mobility tensors in the sense that they express the linear relationship between "action" ($\vec{a}$) and "response" ($\vec{r}$) vectors

$$\vec{r} = \widehat{t}\,\vec{a}.$$

But more advanced tensors are also used to express linear relationships between more simple tensors and vectors. For example, in certain materials mechanical stress can lead to the separation of electric charges and thus create voltage drop between different points of the body. This phenomenon is known as the *piezoelectric effect*. Now if we characterize induced charge separation using a vector $\vec{p} = p^i$, then we can write using index notation

$$p^i = d^{ijk}\sigma_{jk},$$

where σ_{jk} are the components of the stress tensor described above, and d^{ijk} – piezoelectric tensor of the third rank (three indices!)

For readers interested in more examples and details, the book *Physical Properties of Crystals: Their Representation by Tensors and Matrices* by J. F. Ney is highly recommended.

As the last example of tensors in physics, we will consider a more fundamental case from field theory.

6.1.3 Electromagnetic Tensor

In applied physics and engineering one works with electric and magnetic fields that are described using two *different* physical vector quantities: $\vec{E}$ – for electric field strength and $\vec{B}$ – for magnetic field strength.

When a charged particle, say an electron, is placed in an electric field, the latter acts on that particle with a force proportional to the field strength:

$$F_e = qE,$$

where q is the charge of the particle, F_e denotes the force due to the electric field $\vec{E}$.

When the same charged particle is *moving* in a magnetic field, the latter acts on the particle with a force proportional to the magnetic field strength:

$$F_m = qvB,$$

where v is the speed of the charged particle, F_m denotes the force due to the magnetic field $\vec{B}$. The difference between the effects of electric and magnetic fields on a charged particle is illustrated in Figure 6.7.

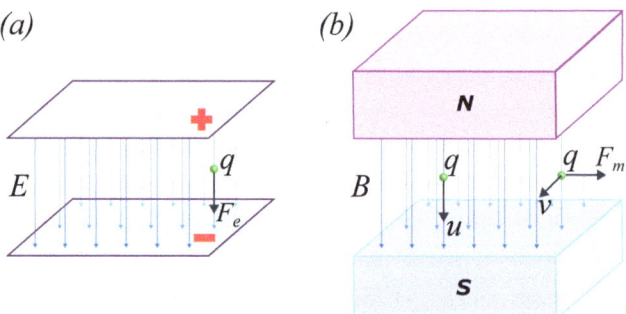

Figure 6.7 Electromagnetic field acts on stationary and moving charged particles. (a) Electric field between the plates of a charged capacitor acts on any charged particle with the force $F_e = qE$ directed along the electric field vector $\vec{E}$. (b) A charged particle moving with the speed v across magnetic field between the poles of a magnet will be affected by a force $F_m = qvB$. The force depends on the relative orientation of the velocity vector $\vec{v}$ and the magnetic field $\vec{B}$. When $\vec{v}$ is parallel to $\vec{B}$ – the force is zero. The force is maximal when $\vec{v}$ is perpendicular to $\vec{B}$.

The distinction between electric and magnetic fields is technical and *not fundamental*. Figuratively speaking, the electric field differs from magnetic field to the same degree as the rest differs from uniform motion. Electric and magnetic fields are different aspects of the same physical entity – *electromagnetic field*.

From the expressions for the electric and magnetic forces F_e and F_m we can see that physical quantities E and B have different units of measurement – a fact which upsets some physicists. They note: If electric and magnetic fields are different aspects of the *same physical object*, they must be measured using the same units, similar to how we measure the height and width of a building using the same units of length.

The way to fix the issue with different units for electric and magnetic fields is to change the way we measure...*velocity*! Nature provides us with a special standard of speed – the speed of light in a vacuum, denoted as c. The speed of light in a vacuum is a "nature-made" absolute quantity, in contrast to such human-made standards as units of length (meter) or time (second). This is why in fundamental physical theories, including the theory of electromagnetic field, it is wise to specify all speeds as fractions of the speed of light.

Thus, in physical formulas, instead of writing v as meters per second, we should use a "normalized" quantity:

$$\bar{v} = v/c \quad \rightarrow \quad v = \bar{v}c.$$

Once we apply this approach to the expression of the magnetic force, we obtain

$$F_m = qvB = q\bar{v}(cB).$$

Now we can see that the quantities E and cB have the same physical meaning – *the force per unit charge*. It is these physical quantities that should be used to describe different aspects of the same electromagnetic field. We will denote them as follows:

$$\vec{E} = \vec{\mathcal{E}} = \mathcal{E}^i, \quad c\vec{B} = \vec{\mathcal{B}} = \mathcal{B}^j.$$

Electromagnetic Tensor

A deep and beautiful discovery of the theory of electromagnetic field can now be stated: Electric and magnetic fields $\mathcal{E}^i$ and $\mathcal{B}^j$ are not separate *vector* quantities, they are, in fact, represent certain *components of a tensor* that describes electromagnetic field.

This tensor is conventionally written as $F^{\mu\nu}$ (F here stands for *field*, not force!) $F^{\mu\nu}$ is a second-rank tensor. The indices μ and ν run from 0 to 3. The relationships between the "usual" electric field and the electromagnetic tensor are given by

$$\mathcal{E}^i = F^{i0}, \quad i = 1, 2, 3.$$

The relationships between the "usual" magnetic field and the electromagnetic tensor can be written in the following way:

$$\mathcal{B}^1 = F^{32}, \mathcal{B}^2 = F^{13}, \mathcal{B}^3 = F^{21}.$$

A convenient way to write all components of a second-rank tensor is to use a table-like structure called *matrix*:

$$F^{\mu\nu} = \begin{pmatrix} F^{00} & F^{01} & F^{02} & F^{03} \\ F^{10} & F^{11} & F^{12} & F^{13} \\ F^{20} & F^{21} & F^{22} & F^{23} \\ F^{30} & F^{31} & F^{32} & F^{33} \end{pmatrix}.$$

In the matrix, the first index μ of $F^{\mu\nu}$ corresponds to the row, while the second index ν corresponds to the column. Both rows and columns are enumerated from 0 to 3.

Using matrix form, we can write the electromagnetic tensor in terms of the electric and magnetic fields:

$$F^{\mu\nu} = \begin{pmatrix} 0 & -\mathcal{E}^1 & -\mathcal{E}^2 & -\mathcal{E}^3 \\ \mathcal{E}^1 & 0 & -\mathcal{B}^3 & \mathcal{B}^2 \\ \mathcal{E}^2 & \mathcal{B}^3 & 0 & -\mathcal{B}^1 \\ \mathcal{E}^3 & -\mathcal{B}^2 & \mathcal{B}^1 & 0 \end{pmatrix}.$$

The last expression makes apparent two features of electromagnetic tensor components. First, all diagonal elements vanish:

$$F^{00} = F^{11} = F^{22} = F^{33} = 0.$$

Second,

$$F^{\mu\nu} = -F^{\nu\mu},$$

a property known as *antisymmetry*. This property requires all diagonal elements to be equal to zero.

Electromagnetic Tensor Components

The first kind of tensor of the second rank that we encountered was a linear operator $\widehat{L}$. The components of any linear operator are given relative to some basis and the components specify how the operator transforms basis vectors:

$$\widehat{L}\,\vec{e}_i = L_{ij}\vec{e}_j.$$

What is the basis used to express components of electromagnetic tensor $F^{\mu\nu}$?

An Electromagnetic tensor is a physical operator which is used to express the action of an electromagnetic field on a moving charged particle. Components of electromagnetic tensor connect special versions of velocity (v_ν) and force (f^μ) acting on a charged particle:

$$f^\mu = qF^{\mu\nu}v_\nu.$$

Without going into details, we will note that on the left-hand side of this equation, we have a four-component force, and on the right-hand side, we have both electric and magnetic effects combined in a single tensor.

6.2   COMPOUND NUMBERS

The machinery of arrow-vectors and operators, which we developed in the previous chapters, helps us establish the connection between planar arrow-vectors and *complex numbers*. We won't be discussing complex numbers in a conventional way, so we will intentionally avoid using conventional name for them, using the term *compound numbers* instead.

A point in a plane is uniquely specified by a pair of Cartesian coordinates (x, y). Alternatively, the point can be specified by an arrow-vector connecting the point and the origin of the coordinates, as shown in the Figure 6.8. If we choose the basis vectors $\vec{e}_1$ and $\vec{e}_2$ along the coordinate axes x and y, respectively, then we can write

$$\vec{a} = x\vec{e}_1 + y\vec{e}_2. \tag{6.2}$$

Thus, there exists a natural connection between planar arrow-vectors and pairs of numbers (x, y):

$$\vec{a} \quad \leftrightarrow \quad (x, y).$$

Any vector $\vec{a}$ can be obtained from the basis vector $\vec{e}_1$ by appropriate scaling and rotation:

$$\vec{a} = a\widehat{R}_\theta\,\vec{e}_1.$$

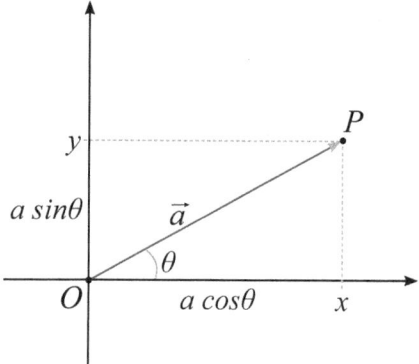

Figure 6.8 A point P in a plane can be specified by an arrow-vector $\vec{a}$, or using a pair of numbers (x, y) – Cartesian coordinates of the point P.

In particular, the second basis vector $\vec{e}_2$ is given by:

$$\vec{e}_2 = \widehat{R}_{\pi/2} \, \vec{e}_1 = \widehat{J} \, \vec{e}_1,$$

where we used a special symbol $\widehat{J}$ for the operator of 90-degree counter-clockwise rotation. The operator $\widehat{J}$ has an interesting property: Applied twice to any vector, it flips the direction of the latter:

$$\widehat{J}\left(\widehat{J}\vec{a}\right) = \left(\widehat{J} \circ \widehat{J}\right)\vec{a} = \widehat{J}^2\,\vec{a} = -\vec{a}.$$

Symbolically this can also be written in the argument-free form:

$$\widehat{J}^2 = -\widehat{I}.$$

Using the operator $\widehat{J}$, we can rewrite the expression (6.2) as follows:

$$\vec{a} = \left(x\widehat{I} + y\widehat{J}\right)\vec{e}_1.$$

Usually, the operator $\widehat{I}$ is omitted if it is clear from the context that expressions involve operators. With this in mind, the expression for the vector $\vec{a}$ can be rewritten:

$$\vec{a} = \left(x + y\widehat{J}\right)\vec{e}_1.$$

Given that $x = a\cos\theta$, and $y = a\sin\theta$, we obtain

$$\vec{a} = a\left(\cos\theta + \sin\theta\,\widehat{J}\right)\vec{e}_1,$$

from which follows the expression for the rotation operator

$$\widehat{R}_\theta = \cos\theta\widehat{I} + \sin\theta\widehat{J} = \cos\theta + \sin\theta\widehat{J}. \tag{6.3}$$

Rotation of a vector by the angle θ can be performed as a single step, or as a sequence of N rotations, each by a smaller step $\delta\theta = \theta/N$. Symbolically, this can be written as a sequence (*composition*) of N identical operations:

$$\widehat{R}_\theta = \widehat{R}_{\delta\theta} \circ \widehat{R}_{\delta\theta} \circ \ldots \circ \widehat{R}_{\delta\theta} = [\widehat{R}_{\delta\theta}]^N.$$

When a vector $\vec{a}$ is rotated by a tiny angle ϕ, its tip travels along the arc with the length $a\phi$, in the direction perpendicular to the vector itself, as illustrated in Figure 6.9. The unit length vector, pointing perpendicular to $\vec{a}$ can be obtained from a unit vector $\vec{u}_a$ pointing along $\vec{a}$ by rotating the former with the operator $\widehat{J}$:

$$\widehat{J}\vec{u}_a = \widehat{J}\left(\frac{\vec{a}}{a}\right).$$

As the result of the rotation by a tiny angle ϕ, the vector $\vec{a}$ becomes

$$\widehat{R}_\phi\vec{a} \approx \vec{a} + a\phi(\widehat{J}\vec{u}_a) = \vec{a} + \phi\widehat{J}\vec{a} = (1 + \phi\widehat{J})\vec{a}.$$

For a tiny angle $\phi = \delta\theta$ this becomes

$$\widehat{R}_{\delta\theta}\vec{a} = (1 + \delta\theta\widehat{J})\vec{a}.$$

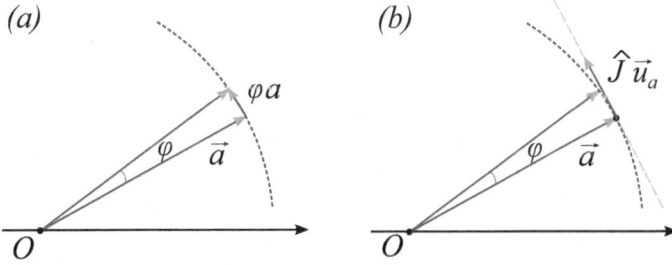

Figure 6.9 Rotation of a vector $\vec{a}$ by a tiny angle ϕ. (a) The tip of the vector $\vec{a}$ will move along the arc of the circle with radius a. The length of the arc is φa. (b) The tangent line to the arc will be perpendicular to the vector $\vec{a}$. The unit vector parallel to the tangent can be found by rotating the unit vector $\vec{u}_a$ with the operator $\widehat{J}$.

Applying this to the relation

$$\widehat{R}_\theta \approx [\widehat{R}_{\delta\theta}]^N$$

we arrive at

$$\widehat{R}_\theta \approx [1 + \delta\theta\,\widehat{J}]^N = [1 + \frac{\theta\widehat{J}}{N}]^N.$$

In the limit of ever-smaller steps ($N \to \infty$ and $\delta\theta \to 0$), the result becomes the famous limit

$$\left(1 + \frac{x}{N}\right)^N = e^x \quad \text{as } N \to \infty.$$

We thus discovered the following expression for the rotation operator $\widehat{R}_\theta$:

$$\widehat{R}_\theta = e^{\theta\widehat{J}}.$$

Comparing this to the expression (6.3), we find the important relation between two different representations of the rotation operator $\widehat{R}_\theta$:

$$\boxed{e^{\theta\widehat{J}} = \cos\theta + \widehat{J}\sin\theta.} \tag{6.4}$$

We can recognize the famous *Euler's formula* in this expression. In words, it expresses the fact that a vector in a plane can be rotated in a single step by the angle θ or using the sequence of tiny rotations by an angle $\delta\theta = \theta/N$ when the number of steps N increases indefinitely.

For the special case of $\theta = \pi$, Euler's formula becomes

$$\boxed{e^{\pi\widehat{J}} = -1 \quad \longrightarrow \quad e^{\pi\widehat{J}} + 1 = 0.} \tag{6.5}$$

So far we have established connections between different ways to represent points in a plane:

1. Coordinate pair (x, y).

2. Arrow-vector $\vec{a}$.

3. Operator sum $z(x, y) = x + y\widehat{J}$.

4. Operator product $z(a, \theta) = ae^{\theta\widehat{J}}$.

Neither of these representations is exceptionally advantageous. Different problems may benefit from different approaches.

We will call the representations $z = x + y\widehat{J}$ and $z = ae^{\theta\widehat{J}}$ *compound numbers*. Compound numbers provide another approach to problems that involve points in a plane. Anything that can be analyzed and solved using either Cartesian coordinates (x, y), or arrow-vectors $\vec{a}$ can also be treated using compound numbers. The apparatus of compound numbers (or *complex numbers*, if you prefer) is extremely powerful and useful in physics and engineering.

Which form of compound numbers works best: sum $z = x + \widehat{J}y$ or product $z = ae^{\theta\widehat{J}}$? The answer depends on the type of operation we want to perform. Given a pair of compound numbers

$$z_a = x_a + y_a\widehat{J} = ae^{\theta_a\widehat{J}}$$

and

$$z_b = x_b + y_b\widehat{J} = be^{\theta_b\widehat{J}},$$

finding their sum $z_c = z_a + z_b$ is easier using sum-form:

$$z_c = (x_a + x_b) + (y_a + y_b)\widehat{J}.$$

On the other hand, the product of two compound numbers $z_c = z_a z_b$ is calculated easily using their product form:

$$z_c = abe^{\theta_a\widehat{J}}e^{\theta_b\widehat{J}}.$$

The last expression can be simplified if we use the trigonometric identity

$$\cos\theta\cos\phi - \sin\theta\sin\phi = \cos(\theta + \phi)$$

and the property $\widehat{J}^2 = -1$:

$$z_c = ab(\cos\theta_a + \sin\theta_a\widehat{J})(\cos\theta_a + \sin\theta_a\widehat{J}) = ab[\cos(\theta_a + \theta_b) + \sin(\theta_a + \theta_b)\widehat{J}].$$

We deduce that

$$z_c = abe^{\theta_a\widehat{J}}e^{\theta_b\widehat{J}} = abe^{(\theta_a + \theta_b)\widehat{J}}.$$

Since a compound number can be represented by a composition of scaling and rotation, the interpretation of the product of two compound numbers is clear: To multiply a number z_a by a number z_b, we must scale the arrow-vector $\vec{a}$ by a factor b – the length of the arrow-vector $\vec{b}$ – and rotate it by an angle θ_b.

Finally, it is evident that multiplication of compound numbers is commutative:

$$z_c = z_a z_b = z_b z_a.$$

Composition Rule

The property

$$z_c = z_a z_b = ab\, e^{(\theta_a + \theta_b)\widehat{J}}$$

can be deduced from the operator form of compound numbers. Indeed, we could write

$$z_c = z_a z_b = ab\, \widehat{R}_{\theta_a} \widehat{R}_{\theta_b}$$

and notice that a sequence of two rotations can be replaced with a single rotation by the combined angle:

$$z_c = ab\, \widehat{R}_{\theta_a + \theta_b} = ab\, e^{(\theta_a + \theta_b)\widehat{J}}.$$

Pauli Matrices

We made heavy use of the operator $\widehat{J}$, which has the components

$$\widehat{J} = \begin{pmatrix} 0 & 1 \\ -1 & 0 \end{pmatrix}.$$

In quantum physics, when dealing with the spin of elementary particles[4], one encounters similar-looking matrices:

$$\widehat{\sigma}_1 = \begin{pmatrix} 0 & 1 \\ 1 & 0 \end{pmatrix} \quad \text{or} \quad \widehat{\sigma}_3 = \begin{pmatrix} 1 & 0 \\ 0 & -1 \end{pmatrix}.$$

These are called *Pauli matrices*[5] after the physicist Wolfgang Pauli. Readers can easily convince themselves that $\widehat{\sigma}_1$ turns the basis vector $\vec{e}_1$ into $\vec{e}_2$ and vice versa, while the operator $\widehat{\sigma}_3$ flips the direction of the basis vector $\vec{e}_2$, leaving $\vec{e}_1$ intact.

6.3 HAMILTONIAN MECHANICS*

Quantum physics makes heavy use of vectors and operators, as well as compound numbers (or complex numbers). However, these mathematical

---

[4]Spin of a particle is its fundamental characteristic, like its mass or charge. However, unlike mass or charge spin is a tensor quantity.

[5]The second Pauli matrix $\widehat{\sigma}_2$ is essentially proportional to the operator $\widehat{J}$.

tools are not exclusive to quantum physics. Classical physics uses similar mathematical tools. Classical physics also shares many fundamental physical concepts with quantum physics, such as *system, state, state evolution, dynamical equation,* and many more. We will now get acquainted with the basics of *Hamiltonian Mechanics* – a way to study mechanical motion using the ideas of energy and momentum as the starting points (as opposed to forces and velocities of Newtonian Mechanics.)

The reader is assumed to be familiar with the basics of Newtonian Mechanics and with the expressions for simple concepts, such as momentum, kinetic energy, or the energy of a stretched spring. The asterisk after the title of this chapter is a reminder that the material that follows is a bit more advanced than the rest.

6.3.1   System and State

A part of nature that can be clearly isolated and studied is called a *physical system.* An electron, an atom, a molecule, a crystal, a pendulum, a comet, a star – these are examples of physical systems of various degrees of complexity.

Since physics is an experimental science, the study of a physical system – or system for short – is based on *measurements* of different *parameters of the system.* Such measurable parameters are also called *observables.* For example, when a projectile is studied, we can measure its *position* and *momentum* at different moments of time. Position and momentum of a body are observables.

In mechanics, the knowledge of position and momentum of a given system at one moment of time t_0 is sufficient to completely know their value at any later moment of time t. In addition, other useful physical quantities, such as energy or angular momentum, can be expressed through position x and momentum p. Symbolically, we can write this as follows:

$$\text{know } (x_0, p_0) \text{ at } t_0 \quad \longrightarrow \quad \text{know } (x, p) \text{ at } t.$$

State of a system is the collection of observables which is, in a certain sense, *complete and self-sufficient.* In other words, state is "all there is to know" about a system. If the state of a system is known at one moment of time t_0, then we should be able to determine the state at any later (or earlier) moment of time t. In classical mechanics the pair of observables (x, p) defines the state of a mechanical system.

Oscillator

In order to better understand the roles of position and momentum we will study a specific system: An *oscillator*. As an oscillator model, we will consider a body with mass m attached to a spring with a spring constant k, as shown in Figure 6.10.

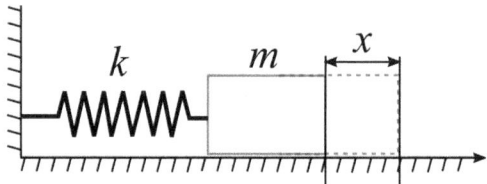

Figure 6.10 Mechanical model of an oscillator: A body with mass m attached to a spring with spring constant k. When displaced from equilibrium by the amount x and released, the oscillator will begin periodic motion.

When the body is moving with speed v, it has kinetic energy

$$E_k = \frac{mv^2}{2}.$$

When the spring is stretched by the amount x, it has potential energy

$$E_p = \frac{kx^2}{2}.$$

The total mechanical energy of the oscillator is given by

$$E = E_k + E_p = \frac{kx^2}{2} + \frac{mv^2}{2}.$$

We can write the total mechanical energy using the momentum $p = mv$ instead of velocity:

$$E = \frac{kx^2}{2} + \frac{p^2}{2m} = H. \tag{6.6}$$

Here we used the letter H to denote the *Hamiltonian* of the system – *the expression of total energy of terms of position and momentum*. In the absence of friction, the total mechanical energy of a system remains constant $(H = \text{const.})$

While both x and p change during the motion of the oscillator, their combination satisfies the relation:

$$H = \frac{x^2}{2/k} + \frac{p^2}{2m} \quad \rightarrow \quad 1 = \frac{x^2}{2H/k} + \frac{p^2}{2mH}.$$

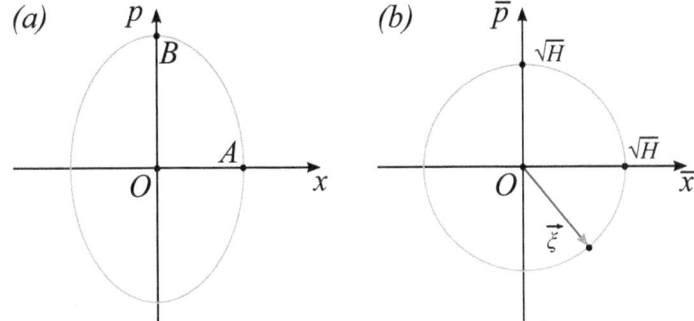

Figure 6.11 (a) During the motion of the oscillator, both the position x and momentum p change, while the point $P = (x, p)$ remains on the ellipsis. (b) The motion of the oscillator in the *normalized* coordinates $\bar{x}$ and $\bar{p}$ is described by a circle with the radius $\xi = \sqrt{\bar{H}}$. A particular combination of $\bar{x}$ and $\bar{p}$ can be described by a *state vector* $\vec{\xi} = \bar{x} + \bar{p}\hat{J}$.

The last expression reminds the equation for an ellipsis with the axes a and b, written for Cartesian coordinates:

$$1 = \frac{x^2}{a^2} + \frac{y^2}{b^2}.$$

Thus, in the xp plane this equation describes an ellipsis, as illustrated in Figure 6.11(a). The maximal value for x can be found from the following condition: At the maximal stretch of the spring, the body stops $(p = 0)$ and all energy is stored in the stretched spring:

$$\frac{kA^2}{2} = H \quad \rightarrow \quad A = \sqrt{\frac{2H}{k}}.$$

When the spring is not stretched and all energy is stored in the moving body, the momentum has the maximal value:

$$\frac{B^2}{2m} = H \quad \rightarrow \quad B = \sqrt{2mH}.$$

The analysis of the oscillator motion can be simplified if we make the equation (6.6) *dimensionless*: Instead of the usual units for distance (m), momentum $(kg * m/s)$, and energy (J), we will express them as scaled versions of some "natural" values x_0, p_0, and H_0:

$$x = \bar{x}x_0, \quad p = \bar{p}p_0, \quad H = \bar{H}H_0,$$

where $\bar{x}$, $\bar{p}$, $\bar{H}$ are unitless quantities. There are many ways to do such a transformation but we will use a simple approach based on the rest-energy of an electron.

Dimensionless Quantities

To understand the idea of dimensionless quantities, consider the following situation: An average human consumes and spends 1800 Calories per day. A physicist might prefer to express this quantity in standard units, and write $E_{day} = 7.531 \times 10^9$ Joules. But we can also express the energy required for a day in terms of the number of ...cookies:

$$E_{day} = 13 E_{ccc} = \bar{E}_{day} E_{ccc}$$

where E_{ccc} is the energy content of a single chocolate chip cookie. In this expression, $\bar{E}_{day}$ is the dimensionless energy.

When an electron is at rest, its energy (*rest-energy*) is given by the most famous equation in physics:

$$H_e = m_e c^2.$$

Now we will use this quantity as the measure of all other energies. We will write $H = \bar{H} H_e$ where $\bar{H}$ is a pure mathematical number (as opposed to "physical" number with units.)

Next we can find the reference value for position. Suppose the oscillator has the energy H_e. Then its maximum displacement will be

$$x_e = \sqrt{\frac{2H_e}{k}},$$

and the maximum momentum

$$p_e = \sqrt{2m H_e}.$$

The Hamiltonian can be written in dimensionless units:

$$\bar{H} H_e = \frac{k x_e^2}{2} \bar{x}^2 + \frac{p_e^2}{2m} \bar{p}^2 = H_e \bar{x}^2 + H_e \bar{p}^2.$$

Therefore, the relationship between the dimensionless position and momentum is described by the equation of a circle in the $\bar{x}\bar{p}$ plane:

$$\bar{H} = \bar{x}^2 + \bar{p}^2.$$

We showed that the state of the oscillator $(\bar{x}, \bar{p})$ can be specified by a point in the $\bar{x}\bar{p}$ plane. The latter can be described by a compound number (a state vector):

$$\vec{\xi} = \bar{x} + \bar{p}\widehat{J}.$$

See Figure 6.11(b) for illustration. The radius of the circle is determined by the total energy:

$$\xi = \sqrt{\bar{H}},$$

where ξ denotes the length of the planar vector $\vec{\xi}$.

6.3.2 State Dynamics

To know the properties of the oscillator at different moments of time we must be able to find its state $\vec{\xi}_t$ at every moment of time, given the initial state $\vec{\xi}_0$ at time $t = 0$. Thus, we must understand what drives the change of the state vector $\vec{\xi}_t$. In other words, we must know the *state dynamics*. The evolution of the state vector $\vec{\xi}$ in time is described by an equation. We will "derive" it now.

As the time advances by a small amount δt, the state vector $\vec{\xi}$ also changes by a small amount $\delta\vec{\xi}$. During motion, both position and momentum change, therefore:

$$\delta\vec{\xi} = \delta\bar{x} + \delta\bar{p}\widehat{J}.$$

On the other hand, since the tip of the state vector follows along the circle of radius ξ, its change can be written as the result of clockwise rotation[6] by a small angle $\delta\phi$:

$$\delta\vec{\xi} = -(\delta\phi\xi)\left[\widehat{J}\left(\frac{\vec{\xi}}{\xi}\right)\right] = -\delta\phi(\widehat{J}\vec{\xi}).$$

If we denote the rate of change of the angle ϕ in time as

$$\omega = \frac{\delta\phi}{\delta t},$$

---

[6]If the oscillator starts from state $\xi_0 = (0, \xi)$ then $\bar{x}$ will grow as $\bar{p}$ decreases.

then we will arrive the equation for the state evolution:

$$\frac{\delta \vec{\xi}}{\delta t} = -\omega \widehat{J} \vec{\xi}. \tag{6.7}$$

Hamiltonian Equations

The left-hand side of (6.7) can be written in terms of the position and momentum:

$$\frac{\delta \vec{\xi}}{\delta t} = \frac{\delta \bar{x}}{\delta t} + \frac{\delta \bar{p}}{\delta t} \widehat{J}.$$

Doing the same with the right-hand side of the equation (6.7), we obtain

$$-\omega \widehat{J} \vec{\xi} = p\omega - \omega x \widehat{J}.$$

We thus derived the time-evolution equations for position and momentum:

$$\frac{\delta \bar{x}}{\delta t} = \omega \bar{p}, \quad \frac{\delta \bar{p}}{\delta t} = -\omega \bar{x}.$$

These equations are called *Hamiltonian equations*; they are the central equations of *Hamiltonian dynamics*. Hamiltonian dynamics is the approach to mechanics alternative to Newtonian dynamics. Instead of using the concept of forces and Newton's second law, Hamiltonian dynamics uses the concept of Hamiltonian – the expression of energy written in terms of position and momentum.

For the oscillator the angular speed ω turns out to be constant. To show this, recall that $\bar{x} = x/x_e$ and $p = \bar{p}p_e$ and write

$$\frac{\delta \bar{x}}{\delta t} = \frac{\delta x}{x_e \delta t} = \frac{v}{x_e}.$$

Here we used the definition of velocity: $v = \delta x/\delta t$.

In Newtonian mechanics the velocity and momentum are related as $p = mv$, therefore

$$\frac{\delta \bar{x}}{\delta t} = \frac{p}{m x_e} = \frac{p_e}{m x_e} \bar{p}.$$

Compare this expression to the first equation of motion to find

$$\omega = \frac{p_e}{m x_e} = \sqrt{\frac{k}{m}},$$

where the last result was obtained after substituting the expressions for x_e and p_e in terms of the reference energy H_e.

Frequency of Oscillations

Note that the expression for the frequency ω of revolution of the state vector $\vec{\xi}$ depends only on the parameters of the oscillator:

$$\omega = \sqrt{\frac{k}{m}}.$$

The choice of units H_e, x_e, and p_e does not affect the value of ω, only the graphical representation of the evolution as circular motion.

Another illuminating relationship exists between the angular speed ω and the energy H_e. To find it we can use the fact

$$\frac{p_e^2}{2m} = H_e \qquad \longrightarrow \qquad \frac{p_e}{m} = \frac{2H_e}{p_e}$$

and arrive at

$$\omega = \frac{2H_e}{x_e p_e}.$$

One last transformation is needed for the equation of state evolution. By acting with the rotation operator $\widehat{J}$ on both sides of

$$\frac{\delta \vec{\xi}}{\delta t} = -\omega \widehat{J} \vec{\xi}$$

we obtain

$$\widehat{J} \frac{\delta \vec{\xi}}{\delta t} = \omega \vec{\xi}.$$

This expression allows us to write the time-evolution equation for the state vector $\vec{\xi}$ in *Schrödinger form*:

$$h_e \widehat{J} \frac{\delta \vec{\xi}}{\delta t} = H_e \vec{\xi}, \tag{6.8}$$

where $h_e = x_e p_e / 2$ – the quantity with the units of angular momentum or, equivalently, energy multiplied by time. Such quantity is known in physics as *action* and it plays an important role in classical and quantum mechanics.

Quantum of Action

The fundamental constant of nature, known as *Planck constant* h, is the smallest possible action. Physicists found that action is quantized: In all physical processes action changes in steps of h.

Schrödinger Equation

The dynamical equation for the oscillator (6.8) is completely analogous to the famous equation of Quantum Physics, known as Schrödinger Equation:

$$i\hbar \frac{\delta \Psi}{\delta t} = H\Psi.$$

Without going into details, we will note that in this equation Ψ corresponds the the state of quantum system, analogous to $\vec{\xi}$. The "imaginary unit" is analogous to the operator $\widehat{J}$ ($i^2 = -1$). The fundamental constant $\hbar$ is the elementary quantum of action. Finally, H represents quantum version of the Hamiltonian.

Having shown that the angular speed ω is constant, we can deduce the time dependence of position and momentum for the oscillator[7]:

$$\bar{x} = \xi \cos \omega t, \quad \bar{p} = \xi \sin \omega t$$

and consequently

$$x(t) = x_e \xi \cos \omega t, \quad p(t) = p_e \xi \sin \omega t.$$

Given that

$$\xi = \sqrt{\frac{H}{H_e}} = \frac{A}{x_e} = \frac{B}{p_e},$$

---

[7]We assume here that initially, at $t = 0$, the oscillator is stretched to the maximum and has no initial momentum.

where A is the maximum displacement of the oscillator with the energy H, and B is the maximum momentum for the same oscillator, we can write

$$x = A\cos\omega t, \quad p = B\sin\omega t.$$

The amplitude of oscillation A and the maximum momentum B can be written in terms of the oscillator parameters k, m, and its total energy H:

$$A = \sqrt{\frac{2H}{k}}, \quad B = \sqrt{2Hm}.$$

This is the mathematical description of the *oscillatory motion*.

6.4 VECTORS AND TENSORS IN COMPUTATION*

As the last example we will consider a somewhat technical problem. This example should demonstrate that the ability to recognize and manipulate tensors might pay off in unexpected places.

Imagine you are exploring a system of tunnels connecting large chambers, as schematically illustrated in Figure 6.12. If we number each entrance/exit (called *ports*) then we can characterize the propagation of sound between a pair (i, j) of ports using two parameters: attenuation of the signal and the time delay between its emission and detection as the signal travels through the structure. In the theory of wave propagation, these two parameters are encoded in so called S-parameter and the set of S-parameters for all combinations $i \rightarrow j$ of ports is called S-matrix.

The relevant theory is developed by Gunnar Filipsson in the paper *A general computer algorithm for S-matrix calculation of interconnected multiports*[8], but here we will only need the central formula from that work:

$$S'_{ij} = S_{ij} + \frac{1}{a_{kl}}[S_{il}S_{kj}(1 - S_{lk}) + S_{ik}S_{lj}(1 - S_{kl}) + \qquad (6.9)$$
$$+S_{il}S_{kk}S_{lj} + S_{ik}S_{ll}S_{kj}],$$

where

$$a_{kl} = 1 - S_{kl} - S_{lk} + S_{kl}S_{lk} - S_{kk}S_{ll}.$$

Note: No summation is implied in expressions like S_{kk} and S_{ll}. The values k and l are fixed and represent the number of ports that connect adjacent

---

[8]https://doi.org/10.1109/EUMA.1981.332972

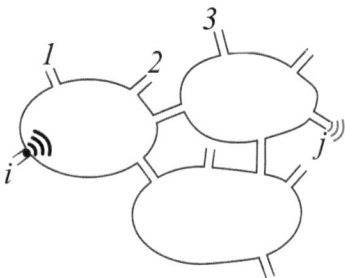

Figure 6.12 A sound signal (acoustic wave) can propagate in a complicated system of chambers connected by tunnels. Shouting at the entrance of one tunnel and listening at the exist of another, we can measure how much delay and attenuation the sound experiences.

chambers. With this in mind, we can introduce special notation in order to get rid of indexed expressions:

$$S_{kk} = \alpha, \quad S_{ll} = \beta, \quad S_{kl} = \mu, \quad S_{lk} = \nu$$

and

$$a_{kl} = a = 1 - \mu - \nu + \mu\nu - \alpha\beta.$$

The problem is then to calculate the set of values S'_{ij} given the set of values S_{ij} for all combinations of port numbers $i, j = 1, 2, \ldots, n$. A simple approach consists of changing the values of each variable $i, j, k,$ and l step by step and performing the usual arithmetic operations at each step. The approach works but it is too slow.

The key to an improved solution is to recognize that the expression for "new" values of S'-parameters can be written in vector-tensor form:

$$\widehat{S}' = \widehat{S} + \widehat{A} + \widehat{B} + \widehat{C} + \widehat{D},$$

where the terms $\widehat{A}$ through $\widehat{D}$ correspond to each term in the parentheses of equation (6.9). For example:

$$A_{ij} = \frac{(1 - \nu)}{a} S_{il} S_{kj}.$$

Here we used the notation introduced above for a and ν.

The next step is to recognize that S_{kj} can be viewed as the j-th component of some vector:

$$S_{kj} = W(k)^j \quad \longrightarrow \quad \overrightarrow{W}(k),$$

while S_{il} is the i-th component of a dual vector:

$$S_{il} = V(l)_i \quad \longrightarrow \quad \overleftarrow{V}(l).$$

Using these vectors, the tensor-operator $\widehat{A}$ can be written using tensor product notation:

$$\widehat{A} = \frac{(1-\nu)}{a} \overleftarrow{V}(l) \otimes \overrightarrow{W}(k).$$

With this approach, we can rewrite the operators $\widehat{B}$, $\widehat{C}$, and $\widehat{D}$ in a similar way:

$$\widehat{B} = \frac{(1-\mu)}{a} \overleftarrow{V}(k) \otimes \overrightarrow{W}(l),$$

$$\widehat{C} = \frac{\alpha}{a} \overleftarrow{V}(l) \otimes \overrightarrow{W}(l),$$

and

$$\widehat{D} = \frac{\beta}{a} \overleftarrow{V}(k) \otimes \overrightarrow{W}(k).$$

The main reason behind these manipulations is that some computations involving vectors and their tensor products can be optimized using advanced algorithms. In the real-world application of this theory, once we transition from the step-by-step approach to the vector-tensor approach, the calculations can be sped up by a factor of 10! The Table 6.1 shows specific calculation time improvement that was made possible after using the vector-tensor nature of the expression. The main point is not that we can save 3 seconds, but that each calculation step is performed 10 times faster. Imagine that a long and complicated calculation that takes 10 days can be done in just one day! That gives the desired result earlier, saves time and also power used by the computer. This all translates into *actual economic benefits*.

TABLE 6.1   Calculations can be done up to 10 times faster when using vector-tensor approach.

Step-by-step	Vector-Tensor
3.0 s	0.276 s

Having learned about vectors and tensors, you may discover new ways to analyze familiar problems in your field of study. Good luck!

CHAPTER HIGHLIGHTS

- *Tensors find application in various areas of science and mathematics.*

- *Geometrical properties of surfaces and spaces can be described using metric tensors.*

- *Physical properties of solids are often anisotropic – depending on the direction of applied "force." Such properties are best described by various tensors: stress tensor, mobility tensor, piezoelectric tensor, and others.*

- *At the fundamental level, electric and magnetic fields are united in a single physical object – electromagnetic field. Electromagnetic field is described by an antisymmetric tensor of the second rank.*

- *The concept of linear operators, and in particular of the rotation operator $\widehat{J}$, can be used to extend the numbers from a number line to the number plane and arrive at complex numbers (or compound numbers, as we called them).*

- *Operators and compound numbers are used in many physical theories, and play an especially important role in Hamiltonian dynamics and quantum mechanics.*

CHAPTER 7

Solutions

DOI: 10.1201/9781003620365-7

Exercise 1.1

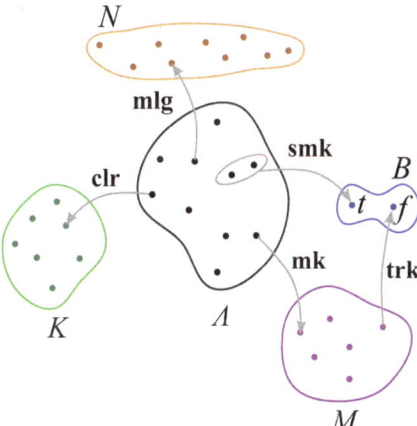

Figure 7.1   The set M contains all possible makes of cars: Ford, Toyota, etc.

The diagram in Figure 7.1 shows the set M – the set of all possible makes of cars. A mapping **trk** returns *true* if a given car maker produces trucks.

Exercise 2.1

Any binary function can be viewed as a unary function if two inputs are replaced by a single input of a *pair of numbers*. Similarly for a function with two outputs. This idea is illustrated in Figure 7.2(a): The function **swp** is viewed as a unary function which swaps the numbers in an *ordered*

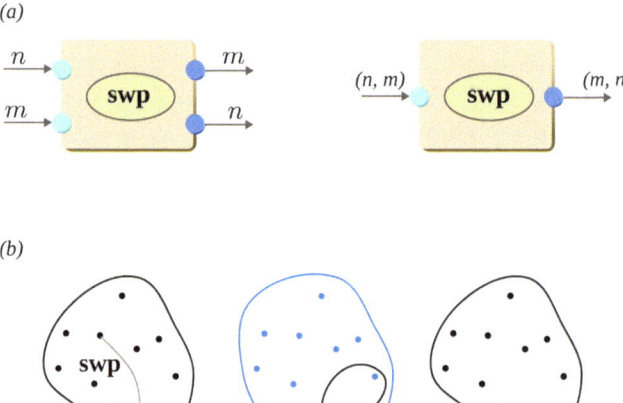

Figure 7.2   (a) Two inputs (outputs) of a function can be replaced with a single input of a *pair* of numbers, turning a binary function into a unary one. (b) Mappings (functions) **swp** and **max** operate on the level of sets: The set of all pairs $\mathbb{Z}^2 = \mathbb{Z} \times \mathbb{Z}$ and the set of triples $\mathbb{Z}^3 = \mathbb{Z} \times \mathbb{Z} \times \mathbb{Z}$. The function **swp** maps the set $\mathbb{Z}^2$ into itself, while the function **max** maps the set $\mathbb{Z}^3$ into $\mathbb{Z}$.

pair:

$$\textbf{swp}\ (n, m) = (m, n).$$

Given the set $\mathbb{Z}$ of whole numbers, we can create the set of all possible *ordered pairs* (n, m). This set can be denoted as follows:

$$(\mathbb{Z}, \mathbb{Z})\ \text{ or }\ \mathbb{Z} \times \mathbb{Z}.$$

The latter notation is standard in mathematics, but the former way of writing is also acceptable. We can similarly denote the set of all *ordered triples*:

$$(\mathbb{Z}, \mathbb{Z}, \mathbb{Z})\ \text{ or }\ \mathbb{Z} \times \mathbb{Z} \times \mathbb{Z}.$$

With the notation introduced above, the action of functions with multiple inputs or outputs can be depicted on the level of sets. Figure 7.2(b) shows how this works for the functions **swp** and **max** .

Exercise 2.2

Consider a binary function that accepts a pair of natural numbers and returns the third natural number in the following way:

$$\textbf{rep}\ 3\,2 = 33 \qquad \textbf{rep}\ 1\,4 = 1111.$$

Thus, the output is a natural number with identical digits given by the first number, repeated a number of times specified by the second number.

Infix variant of this operation can be written, rather arbitrarily, like this:

$$\textbf{rep}\ n\,m = n \succ m = nnn\ldots n.$$

Exercise 2.3

A linear function f must satisfy the linearity condition

$$f\,(a * n) = a * (f\,n).$$

For $a = 0$ we must have

$$f\,(0 * n) = 0 * (f\,n),$$

or, equivalently

$$f\,0 = 0.$$

Also, for $a = m$ and $n = 1$ we must have

$$f\,(m * 1) = m * (f\,1),$$

from which follows

$$f\,m - m(f\,1).$$

Exercise 2.4

The schematics in Figure 7.3(a) and (b) demonstrate the two linearity requirements.

Exercise 2.5

Using Einstein's summation rule, a polynomial of degree n can be written as follows:

$$P_n\,x = a_i x^i, \quad i = 0, 1, 2 \ldots, n.$$

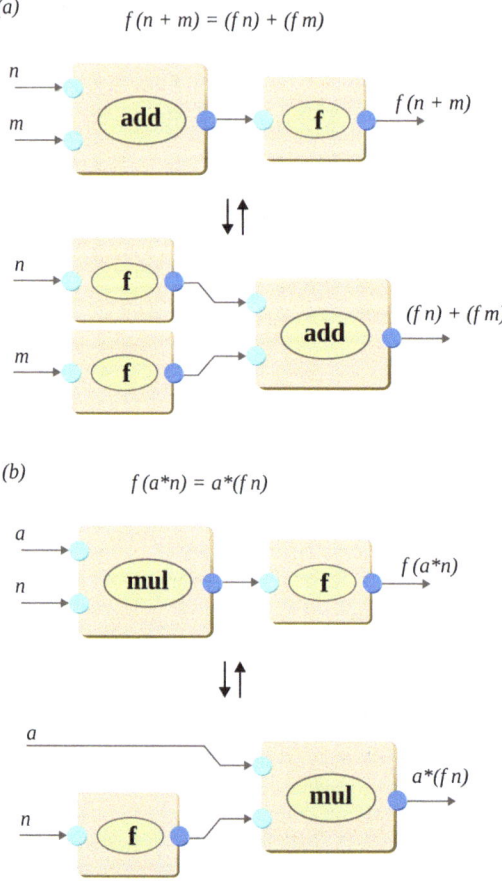

Figure 7.3 The linearity conditions can be represented schematically with different relative configurations (order) of the "boxes."

Exercise 2.6

The expression

$$b_i y_i, \quad i = 1, 2, 3, 4$$

represents the sum of four terms:

$$S = b_1 y_1 + b_2 y_2 + b_3 y_3 + b_4 y_4.$$

Similarly,

$$b_j y_j, \quad j = 1, 2, 3, 4$$

stands for the same sum S, just as the expression

$$b_k y_k = b_1 y_1 + b_2 y_2 + b_3 y_3 + b_4 y_4.$$

Exercise 2.7

In the expression

$$(a_i x_i)(a_j x_j)$$

both parentheses contain the identical sum:

$$a_i x_i = a_j x_j = a_1 x_1 + a_2 x_2.$$

Opening the parentheses, we obtain

$$(a_i x_i)(a_j x_j) = a_1^2 x_1^2 + a_2^2 x_2^2 + 2 a_1 a_2 x_1 x_2.$$

In contrast, the expression $a_i^2 x_i^2$ stands for

$$a_i^2 x_i^2 = a_1^2 x_1^2 + a_2^2 x_2^2.$$

Clearly,

$$a_i^2 x_i^2 \neq (a_i x_i)^2.$$

Exercise 2.8

The left-hand side of the expression

$$(a_i x_i)^2 = \frac{b_j y_j}{c_k c_k}$$

can be written as

$$(a_i x_i)^2 = \left(\sum_{i=1}^{i=N} a_i x_i \right)^2$$

The right-hand side takes the form:

$$\frac{b_j y_j}{c_k c_k} = \frac{\sum\limits_{j=1}^{j=N} b_j y_j}{\sum\limits_{k=1}^{k=N} c_k c_k}.$$

Therefore, the original equality can be re-written using the traditional summation sign:

$$\left(\sum_{i=1}^{i=N} a_i x_i \right)^2 = \frac{\sum\limits_{j=1}^{j=N} b_j y_j}{\sum\limits_{k=1}^{k=N} c_k c_k}.$$

Already one can see the advantage of Einstein's summation rule.

Exercise 2.9

The expression

$$\delta_{1i}a_i, \quad i = 1, 2, 3, 4$$

represents the sum

$$\delta_{11}a_1 + \delta_{12}a_2 + \delta_{13}a_3 + \delta_{14}a_4.$$

The only non-zero term corresponds to $\delta_{11} = 1$, therefore

$$\delta_{1i}a_i = a_1.$$

Similarly, we have

$$\delta_{3k}a_k, \quad k = 1, 2, 3, 4$$

representing

$$\delta_{31}a_1 + \delta_{32}a_2 + \delta_{33}a_3 + \delta_{34}a_4 = \delta_{33}a_3.$$

Consequently,

$$\delta_{3k}a_k = a_3.$$

Next,

$$\epsilon_{1j}a_j = \epsilon_{11}a_1 + \epsilon_{12}a_2 + \epsilon_{13}a_3 + \epsilon_{14}a_4$$

can be simplified to

$$\epsilon_{1j}a_j = a_2 + a_3 + a_4$$

using the definition of ϵ_{ij}.

Finally, the expression

$$\epsilon_{3j}a_j = \epsilon_{31}a_1 + \epsilon_{32}a_2 + \epsilon_{33}a_3 + \epsilon_{34}a_4$$

is reduced to

$$\epsilon_{3j}a_j = a_4 - a_1 - a_2.$$

Exercise 2.10

The sum

$$a_i + a_j$$

can be rewritten using the facts $a_j = \delta_{ji}a_i$ and $\delta_{ij} = \delta_{ji}$:

$$a_i + a_j = a_i + \delta_{ij}a_i = (1 + \delta_{ij})a_i = (1 + \delta_{ji})a_i.$$

Exercise 2.11

(a) The expression $\delta_{ij}a_ib_j$ can be simplified using the fact $\delta_{ij}b_j = b_i$:

$$\delta_{ij}a_ib_j = a_ib_i = a_1b_1 + a_2b_2.$$

(b) Fully writing out $\epsilon_{ij}a_ib_j$ results in

$$\epsilon_{11}a_1b_1 + \epsilon_{12}a_1b_2 + \epsilon_{21}a_2b_1 + \epsilon_{22}a_2b_2.$$

From the definition of ϵ_{ij} follows that only the terms with $i \neq j$ survive:

$$\epsilon_{ij}a_ib_j = a_1b_2 - a_2b_1.$$

Exercise 2.12

(a) Firstly, we can recall that when δ_{ij} is summed with a vector a_j it simply "renames" the index that is being used for summation:

$$\delta_{ij}a_j = a_i.$$

Using this property, we immediately get

$$\delta_{ij}\delta_{jk} = \delta_{ik}.$$

Another – and longer – way to get this result is to write out the summation fully:

$$\delta_{ij}\delta_{jk} = \delta_{i1}\delta_{1k} + \delta_{i2}\delta_{2k} + \ldots + \delta_{in}\delta_{nk}.$$

If $i \neq k$, all terms on the right are zero. Indeed, $\delta_{i1}\delta_{1k}$ is zero unless $i = 1$ and $k = 1$; similarly, $\delta_{i2}\delta_{2k}$ is zero unless $i = 2$ and $k = 2$ and so on. Therefore, the only non-zero value for $\delta_{ij}\delta_{jk}$ is when $i = k$. Let $i = k = m$, then in the sum

$$\delta_{m1}\delta_{1m} + \delta_{m2}\delta_{2m} + \ldots + \delta_{mm}\delta_{mm} + \ldots + \delta_{mn}\delta_{nm}.$$

there is only one non-zero term, namely

$$\delta_{mm}\delta_{mm} = 1 \cdot 1 = 1.$$

Summarizing the above arguments, we conclude that

$$\delta_{ij}\delta_{jk} = 1 \text{ if } i = k \text{ and } 0 \text{ otherwise.}$$

This is equivalent to the expression

$$\delta_{ij}\delta_{jk} = \delta_{ik}.$$

(b) The expression $\epsilon_{ij}\epsilon_{jk}$, when fully expanded as a sum, takes the form

$$\epsilon_{ij}\epsilon_{jk} = \epsilon_{i1}\epsilon_{1k} + \epsilon_{i2}\epsilon_{2k}.$$

If $i = k = 1$, the sum is reduced to

$$\epsilon_{1j}\epsilon_{j1} = \epsilon_{11}\epsilon_{11} + \epsilon_{12}\epsilon_{21} = 1 \cdot (-1) = -1.$$

Similarly, for $i = k = 2$ we get

$$\epsilon_{2j}\epsilon_{j2} = \epsilon_{21}\epsilon_{12} + \epsilon_{22}\epsilon_{22} = (-1) \cdot 1 = -1.$$

On the other hand, if $i = 1$ and $k = 2$, we obtain

$$\epsilon_{1j}\epsilon_{j2} = \epsilon_{11}\epsilon_{12} + \epsilon_{12}\epsilon_{22} = 0.$$

Same for $i = 2$ and $k = 1$:

$$\epsilon_{2j}\epsilon_{j1} = \epsilon_{21}\epsilon_{11} + \epsilon_{22}\epsilon_{21} = 0.$$

We thus manually checked all cases and showed that

$$\epsilon_{ij}\epsilon_{jk} = -\delta_{ik} \quad i, j, k = 1, 2.$$

Exercise 2.13

Let us denote:

$$x = \epsilon_{ij}a_i a_j.$$

Since we can rename the summation indices, we can write

$$\epsilon_{ij}a_i a_j = \epsilon_{ik}a_i a_k = \epsilon_{jk}a_j a_k = \epsilon_{ji}a_j a_i.$$

Now we have $\epsilon_{ji} = -\epsilon_{ij}$ and this leads to

$$\epsilon_{ji}a_j a_i = -\epsilon_{ij}a_i a_j.$$

We thus showed that $x = -x$ and therefore $x = 0$.

Exercise 3.1

An expansion of an arbitrary vector $\vec{a}$ in terms of the basis vectors is given by

$$\vec{a} = a_1\vec{e}_1 + a_2\vec{e}_2 + a_3\vec{e}_3 + \ldots a_n\vec{e}_n.$$

This can be compactly written using Einstein's summation rule:

$$\vec{a} = a_i\vec{e}_i \quad i = 1, 2, \ldots, n.$$

If the number of basis vectors is known and fixed, as is usually the case, we can omit the range of the summation index and simply write

$$\vec{a} = a_i\vec{e}_i.$$

Exercise 3.2

The expression

$$\vec{e}\,'_1 = E_{11}\vec{e}_1 + E_{12}\vec{e}_2,$$

can be written using Einstein's summation rule as follows:

$$\vec{e}\,'_1 = E_{1j}\vec{e}_j.$$

Similarly for the second basis vector:

$$\vec{e}\,'_2 = E_{2j}\vec{e}_j.$$

Combining both results, we obtain

$$\vec{e}\,'_i = E_{ij}\vec{e}_j.$$

Exercise 3.3

(a) Writing the expansion of the "new" basis as follows:

$$\vec{e}\,'_1 = \mu\vec{e}_1 + 0\vec{e}_2,$$

$$\vec{e}\,'_2 = 0\vec{e}_1 + \nu\vec{e}_2,$$

we can immediately find the components E_{ij}:

$$E_{11} = \mu, \ E_{12} = 0, \ E_{21} = 0, \ E_{22} = \nu.$$

We note that the "new" basis vectors are simply scaled versions of the "old" ones: $\vec{e}\,'_i$ is parallel to $\vec{e}_i$ but may have different lengths (if $\mu, \nu \neq 1$).
(b) The simple relations between the "new" and "old" basis vectors allow us to find

$$\vec{e}_1 = \vec{e}\,'_1/\mu$$

and

$$\vec{e}_2 = \vec{e}\,'_2/\nu.$$

If the vector $\vec{a}$ is expanded using the "old" basis:

$$\vec{a} = a_1\vec{e}_1 + a_2\vec{e}_2,$$

then we can write

$$\vec{a} = (a'_1/\mu)\vec{e}\,'_1 + (a'_2/\nu)\vec{e}\,'_2,$$

and immediately find

$$a_1' = a_1/\mu, \quad a_2' = a_2/\nu.$$

Therefore, when the "new" basis vectors are scaled by factors μ and ν, the corresponding "new" components of the vectors are scaled by $1/\mu$ and $1/\nu$ – *in the opposite direction*, to counter the effect of basis variation. The arrow-like vectors are thus called *contravariant vectors*.

Exercise 3.4

The compact expression

$$E_{ij}' E_{jk}$$

for $i = 1$ and $k = 2$ can be expanded into a sum:

$$E_{1j}' E_{j2} = E_{11}' E_{12} + E_{12}' E_{22}.$$

Exercise 3.5

The system of four equations

$$
\begin{array}{rcll}
aw + cx & = & 1, & (7.1) \\
bw + dx & = & 0, & (7.2) \\
ay + cz & = & 0, & (7.3) \\
by + dz & = & 1 & (7.4)
\end{array}
$$

can be solved by noticing that the first two equations do not involve the unknowns from the second pair of equations, and vice versa.

From the equation

$$bw + dx = 0$$

we first find $w = -dx/b$ and substitute it into the first equation:

$$-adx/b + cx = 1,$$

from which we easily find

$$x = \frac{b}{cb - ad} = -\frac{b}{\Delta},$$

where we introduced the notation $\Delta = ad - bc$. Then

$$w = -\frac{dx}{b} = \frac{d}{\Delta}.$$

The second pair of equations can be solved similarly. First, we get

$$z = -\frac{ay}{c},$$

and substitute it into the last of four equations:

$$by - \frac{ady}{c} = 1.$$

From the last expression follows

$$y = -\frac{c}{\Delta}.$$

Consequently,

$$z = \frac{a}{\Delta}.$$

Exercise 3.6

Firstly, we start with the compact expression

$$E_{ij}E'_{jk} = \delta_{ik}$$

and write it out fully for all four combinations of the indices i and k:

$$
\begin{aligned}
E_{11}E'_{11} + E_{12}E'_{21} &= 1, \\
E_{11}E'_{12} + E_{12}E'_{22} &= 0, \\
E_{21}E'_{11} + E_{22}E'_{21} &= 0, \\
E_{21}E'_{12} + E_{22}E'_{22} &= 1.
\end{aligned}
$$

Secondly, using the notation

$$E_{11} = a, \quad E_{12} = b, \quad E_{21} = c, \quad E_{22} = d,$$

and

$$E'_{11} = w, \quad E'_{12} = x, \quad E'_{21} = w, \quad E'_{22} = z,$$

we arrive at the four equations which we can group into two pairs of equations, each pair involving only two unknowns. The first pair is

$$
\begin{aligned}
aw + by &= 1, \\
cw + dy &= 0;
\end{aligned}
$$

the second pair:

$$cx + dz \quad - \quad 1,$$
$$ax + bz \quad = \quad 0.$$

The first pair is easily solved when we find

$$w = -\frac{dy}{c},$$

and substitute it into the first equation of the first pair:

$$-\frac{ady}{c} + by = 1,$$

from which follows:

$$y = -\frac{c}{\Delta} \qquad \Delta = ad - bc.$$

Immediately we get

$$w = \frac{d}{\Delta}.$$

Similarly, we first find

$$z = -\frac{ax}{b},$$

and substitute into the first equation of the second pair:

$$cx - \frac{adx}{b} = 1.$$

Solving for x, we get

$$x = \frac{b}{\Delta},$$

and therefore

$$z = \frac{a}{\Delta}.$$

We conclude that although two conditions $E'_{ij}E_{jk} = \delta_{ik}$ and $E_{ij}E'_{jk} = \delta_{ik}$ result in slightly different equations, they put the same constraints on the relations between the coefficients $E_{ij}(a, b, c, d)$ and $E'_{nm}(w, x, y, z)$.

Exercise 4.1

The equation of a circle with the radius R can be written using Cartesian coordinates:

$$x^2 + y^2 = R^2.$$

The transformation

$$b_1 = a_1 + a_2, \quad b_2 = a_1 * a_2$$

moves every point (x, y) into a new point (x', y') related by the same equations:

$$x' = x + y, \quad y' = xy.$$

Squaring x', we get

$$(x')^2 = x^2 + y^2 + 2xy = R^2 + 2y'.$$

Therefore, the components of the transformed vector are related as follows:

$$y' = (x')^2/2 - R^2/2 \quad \Leftrightarrow \quad b_2 = b_1^2/2 - R^2/2.$$

Exercise 4.2

The operator of the normalization $\widehat{N}$ fails to satisfy the first linearity condition because

$$\widehat{N}(\alpha\vec{a}) \neq \alpha(\widehat{N}\vec{a}).$$

Indeed, the left-hand side must be a unit vector in the direction of $\alpha\vec{a}$, which is the same as the direction of $\vec{a}$:

$$\widehat{N}(\alpha\vec{a}) = \vec{u}_a = \widehat{N}\vec{a}.$$

In addition, the operator $\widehat{N}$ does not satisfy the second linearity condition:

$$\widehat{N}(\vec{a} + \vec{b}) = (\widehat{N}\vec{a}) + (\widehat{N}\vec{b}).$$

Take, for instance, $\vec{a} = \vec{e}_1$ and $\vec{b} = 1000\vec{e}_2$. The sum-vector $\vec{a} + \vec{b}$ will be pointing almost along the second basis vector $\vec{e}_2$, therefore

$$\widehat{N}(\vec{e}_1 + 1000\vec{e}_2)$$

will be a unit vector *almost parallel* to $\vec{e}_2$. However, the vector

$$(\widehat{N}\vec{e}_1) + (\widehat{N}[1000\vec{e}_2]) = \vec{e}_1 + \vec{e}_2$$

will go diagonally between $\vec{e}_1$ and $\vec{e}_2$.

Exercise 4.3

The condition for the degeneracy of a linear operator can be written as follows:

$$\widehat{L}\vec{e}_1 = \lambda\left(\widehat{L}\vec{e}_2\right).$$

For simplicity, let us denote $\vec{v} = \widehat{L}\vec{e}_2$. Then we have

$$\widehat{L}\vec{e}_1 = \lambda\vec{v}.$$

For an arbitrary vector $\vec{a}$ we can find the action of the degenerate linear operator $\widehat{L}$

$$\widehat{L}\vec{a} = \widehat{L}(a_1\vec{e}_1 + a_2\vec{e}_2) = a_1(\widehat{L}\vec{e}_1) + a_2(\widehat{L}\vec{e}_2).$$

Now we can use the degeneracy condition to find

$$\widehat{L}\vec{a} = a_1\lambda\vec{v} + a_2\vec{v} = (a_1\lambda + a_2)\vec{v} = \alpha\vec{v}.$$

Thus, any vector $\vec{a}$ is mapped into a vector parallel to $\vec{v}$. In other words, the degenerate linear operator "collapses" all vectors onto a single line.

Exercise 4.4

The relation

$$\widehat{L}\vec{a} = \vec{b}$$

can be written using components relative to the "new" basis $\{\vec{e}\,'_i\}$. Expanding the vector $\vec{a}$, we get:

$$\widehat{L}(a'_i\vec{e}\,'_i) = a'_i(\widehat{L}\vec{e}\,'_i).$$

Now the components of the operator $\widehat{L}$ relative to the "new" basis are defined similarly to the components relative to the "old" basis:

$$\widehat{L}\vec{e}\,'_i = L'_{ij}\vec{e}\,'_j.$$

Combining the last two expressions, we obtain

$$\widehat{L}\vec{a} = (a'_i L'_{ij})\vec{e}\,'_j,$$

where we indicated that the summation with the components of the vector $\vec{a}$ happens first. Comparing this result with the expansion of the vector $\vec{b}$ relative to the "new" basis

$$\vec{b} = b'_j\vec{e}\,'_j,$$

we can see that the following relation holds

$$a'_i L'_{ij} = b'_j.$$

Exercise 4.5

To evaluate the right-hand side of the expression

$$L_{11} + L_{22} = L'_{11} + L'_{22},$$

we need to recall to rule of transformation of components of the linear operator:

$$L'_{mj} = E_{mk} L_{ki} E'_{ij}.$$

Using the last expression we can write

$$L'_{11} = E_{1k} L_{ki} E'_{i1} = (E'_{i1} E_{1k}) L_{ki}$$

and

$$L'_{22} = E_{2k} L_{ki} E'_{i2} = (E'_{i2} E_{2k}) L_{ki}.$$

Summing up, we obtain

$$L'_{11} + L'_{22} = (E'_{i1} E_{1k} + E'_{i2} E_{2k}) L_{ki}.$$

The sum in the parentheses can be made more compact using Einstein's summation rule:

$$E'_{i1} E_{1k} + E'_{i2} E_{2k} = E'_{ij} E_{jk}.$$

We showed that

$$E'_{ij} E_{jk} = \delta_{ik},$$

therefore

$$L'_{11} + L'_{22} = \delta_{ik} L_{ki} = L_{ii} = L_{11} + L_{22}.$$

Exercise 5.1

The operator $\angle$ is not linear in either of its arguments. Indeed, scaling the first argument by an arbitrary factor α does not affect the measured angle:

$$\angle (\alpha \vec{a}) \, \vec{b} = \angle \, \vec{a} \, \vec{b} \neq \alpha (\angle \, \vec{a} \, \vec{b}).$$

Same applies to the second argument.

Exercise 5.2

Components of any linear operator are defined as the coefficients in the expansion

$$\widehat{L}\,\vec{e}_i = L_{ij}\vec{e}_j.$$

For the potential operator $\widehat{\beta}$ this means

$$\widehat{\beta}\,\vec{e}_i = (a_i b_j)\vec{e}_j = a_i(b_j\vec{e}_j) = a_i\vec{b}.$$

Thus, the operator $\widehat{\beta}$ maps all basis vectors into vectors parallel to $\vec{b} = b_j\vec{e}_j$. Consequently, the operator $\widehat{\beta}$ maps *all* vectors into the same direction parallel to the vector $\vec{b}$. It is an example of a *degenerate linear operator*. See also Exercise 4.3.

Exercise 5.3

By definition

$$\overleftarrow{a} = \widehat{\sigma}\,\vec{a} = \widehat{\sigma}\,(a_1\vec{e}_1 + a_2\vec{e}_2).$$

Using the linearity of $\widehat{\sigma}$, we first write

$$\widehat{\sigma}\,(a_1\vec{e}_1 + a_2\vec{e}_2) = [\widehat{\sigma}\,(a_1\vec{e}_1)] + [\widehat{\sigma}\,(a_2\vec{e}_2)].$$

Using the linearity again, we get

$$\widehat{\sigma}\,(a_1\vec{e}_1) = a_1(\widehat{\sigma}\,\vec{e}_1) = a_1\overleftarrow{e}_1$$

and

$$\widehat{\sigma}\,(a_2\vec{e}_2) = a_2(\widehat{\sigma}\,\vec{e}_2) = a_2\overleftarrow{e}_2,$$

which lead to

$$\overleftarrow{a} = \widehat{\sigma}\,(a_1\vec{e}_1 + a_2\vec{e}_2) = a_i\overleftarrow{e}_i.$$

We showed that the vector conjugate to $\vec{a}$ can be expanded in terms of basis conjugate to $\vec{e}_i$.

Exercise 5.4

The determinant of a linear operator $\widehat{L}$ can be calculated from its components according to:

$$det\,\widehat{L} = L_{11}L_{22} - L_{12}L_{21}.$$

For a projector $\widehat{\underline{\underline{A}}}$ we have

$$\underline{\underline{A}}_{11}\,\underline{\underline{A}}_{22} = \frac{a_1 a_1 a_2 a_2}{a^4} = \frac{a_1^2 a_2^2}{a^4}$$

and

$$\underline{\underline{A}}_{12}\,\underline{\underline{A}}_{21} = \frac{a_1 a_2 a_2 a_1}{a^4} = \frac{a_1^2 a_2^2}{a^4}.$$

It immediately follows that $det\,\widehat{L} = 0$.

A helpful related exercise is Exercise 5.2.

Exercise 5.5

(a) First, we can write symbolically:

$$\widehat{\underline{\underline{A}}} \circ \widehat{\underline{\underline{A}}} = \frac{(\vec{a}\,\overleftarrow{a})}{a^2}\,\frac{(\vec{a}\,\overleftarrow{a})}{a^2} = \frac{\vec{a}\,(\overleftarrow{a}\,\vec{a})\,\overleftarrow{a}}{a^4},$$

which, using the fact $\overleftarrow{a}\,\vec{a} = a^2$ is reduced to

$$\widehat{\underline{\underline{A}}} \circ \widehat{\underline{\underline{A}}} = \frac{\vec{a}\,\overleftarrow{a}}{a^2} = \widehat{\underline{\underline{A}}}.$$

Second, using components, we write the product of two operators as follows:

$$(\underline{\underline{A}}_{ij})(\underline{\underline{A}}_{jk}) = \frac{a_i a_j a_j a_k}{a^4}.$$

Recalling that $a_j a_j = a^2$, we find

$$(\underline{\underline{A}}_{ij})(\underline{\underline{A}}_{jk}) = \frac{a_i a_k}{a^2} = \underline{\underline{A}}_{ik}.$$

Every projector of the type $L = \vec{d}\,\overleftarrow{d}/d^2$ has this property.

(b) For a composition of two projectors

$$\widehat{L} = \widehat{\underline{\underline{B}}} \circ \widehat{\underline{\underline{A}}}$$

the components are given by

$$L_{ik} = \lambda\,a_i b_k, \qquad \lambda = \frac{\vec{a}\cdot\vec{b}}{a^2 b^2}.$$

Composition of two such operators can be evaluated using their components:

$$L_{ik}L_{kj} = (\lambda\, a_i b_k)(\lambda\, a_k b_j) = \lambda^2 a_i(a_k b_k)b_j = \lambda^2(\vec{a} \cdot \vec{b})a_i b_j.$$

We thus showed that

$$\widehat{L} \circ \widehat{L} = \lambda(\vec{a} \cdot \vec{b})\widehat{L}.$$

Now

$$\lambda(\vec{a} \cdot \vec{b}) = \frac{(\vec{a} \cdot \vec{b})^2}{a^2 b^2} = \cos\theta,$$

where θ is the angle between the vectors $\vec{a}$ and $\vec{b}$. Therefore,

$$\widehat{L} \circ \widehat{L} = \cos\theta\widehat{L} \neq \widehat{L}.$$

Only for $\vec{a} = \vec{b}$ we have the property $\widehat{L} \circ \widehat{L} = \widehat{L}$.

Exercise 5.6

The components of the composition

$$\widehat{L} = \underline{\underline{\widehat{B}}} \circ \underline{\underline{\widehat{A}}}.$$

are

$$L_{ik} = \lambda\, a_i b_k, \quad \lambda = \frac{\vec{a} \cdot \vec{b}}{a^2 b^2}.$$

Reversing the order of arguments of the composition results in

$$\widehat{M} = \underline{\underline{\widehat{A}}} \circ \underline{\underline{\widehat{B}}},$$

with the components

$$M_{ik} = \lambda\, b_i a_k, \quad \lambda = \frac{\vec{b} \cdot \vec{a}}{a^2 b^2}.$$

In general, $M_{12} \neq L_{12}$ because $a_1 b_2 \neq b_1 a_2$.

Exercise 5.7

To arrive at the transformation rules for the components of different types of tensors, we will use a simple fact: A general tensor with components t^{ij} will behave just like the tensor product of two contra-variant vectors $a^i b^j$. Thus, we will study four types of tensor products:

$$a^i b^j, \quad a_i b_j, \quad a^i b_j, \quad a_i b^j.$$

It will be helpful to recall the transformation rules of contravariant and covariant vectors. An arrow-like contravariant vector can be expanded in a basis:

$$\vec{a} = a^i \vec{e}_i.$$

Every vectors from a "new" basis can be similarly expanded:

$$\vec{e}'_j = E_{j\,\bullet}^{\bullet\,k} \vec{e}_k.$$

Here we deliberately included the "dummy" symbol " $\bullet$ " to visually align the indices according to their order. Expanding the same vector $\vec{a}$ relative to the "new" basis has the form:

$$\vec{a} = a'^i \vec{e}'_i,$$

and similarly

$$\vec{a} = a^i \vec{e}_i.$$

$$\vec{e}_k = (E')_{k\,\bullet}^{\bullet\,j} \vec{e}_j.$$

We also obtained the relation between the components of the same vector $\vec{a}$ in different bases:

$$a'^i = (E')_{k\,\bullet}^{\bullet\,i} a^k.$$

Another contravariant vector will have similar relations:

$$b'^j = (E')_{l\,\bullet}^{\bullet\,j} b^l,$$

and their tensor product $t^{ij} = a^i b^j$ will be transformed according to

$$(t')^{ij} = (E')_{k\,\bullet}^{\bullet\,i} (E')_{l\,\bullet}^{\bullet\,j} t^{kl}.$$

Using the transformation rule of covariant components:

$$a'_i = E_{i\,\bullet}^{\bullet\,k} a_k \quad \text{and} \quad b'_j = E_{j\,\bullet}^{\bullet\,l} b_l,$$

we immediately arrive at transformation rule of the components of a doubly-covariant tensor:

$$t'_{ij} = E_{i\bullet}^{\bullet k} E_{j\bullet}^{\bullet l} t_{kl}.$$

In a similar fashion, by combining the transformation rules of contravariant and covariant vectors, we can obtain the transformation of contra-covariant tensor:

$$(t')_{\bullet j}^{i\bullet} = (E')_{k\bullet}^{\bullet i} E_{j\bullet}^{\bullet l} t_{\bullet l}^{k\bullet},$$

and covariant-contravariant tensor:

$$(t')_{i\bullet}^{\bullet j} = E_{i\bullet}^{\bullet k} (E')_{l\bullet}^{\bullet j} t_{k\bullet}^{\bullet l}.$$

Exercise 6.1

(a) The action of the metric tensor on a tensor product $\vec{a} \otimes \vec{b}$ can be understood once we write it out using components. First, let's write the expression for the distance squared:

$$d^2 = \eta_{ij} d^i d^j.$$

In a similar way, we can write

$$\widehat{\eta}(\vec{a} \otimes \vec{b}) = \eta_{ij} a^i b^j.$$

Using the fact

$$\eta_{ij} = \widehat{\sigma} \, \vec{e}_i \, \vec{e}_j$$

and bilinear nature of the dol-operator $\widehat{\sigma}$, we deduce

$$\eta_{ij} a^i b^j = a^i b^j (\widehat{\sigma} \, \vec{e}_i \, \vec{e}_j) = \widehat{\sigma} \, (a^i \vec{e}_i)(b^j \vec{e}_j) = \widehat{\sigma} \vec{a} \, \vec{b}.$$

Thus, we showed that

$$\widehat{\eta}(\vec{a} \otimes \vec{b}) = \vec{a} \cdot \vec{b}.$$

The connection between the metric tensor and scalar products is not surprising. Indeed, if we recall that if vector $\vec{d}$ connects two points in a plane and is given by

$$\vec{d} = \vec{b} - \vec{a}, \quad d^i = b^i - a^i,$$

then

$$\eta_{ij} d^i d^j = \eta_{ij} (a^i - b^i)(a^j - b^j) = \eta_{ij} a^i a^j + \eta_{ij} b^i b^j + 2\eta_{ij} a^i b^j.$$

We arrived at the familiar theorem of planar geometry – theorem of cosine:

$$d^2 = a^2 + b^2 - 2ab\cos\theta.$$

(b) Vectors of orthonormal basis all have unit lengths (*normalized* vectors):

$$e_i^2 = \widehat{\sigma}\,\vec{e}_i\,\vec{e}_i = \eta_{ii} = 1.$$

Different vectors of orthonormal basis are perpendicular to each other (*orthogonal* vectors):

$$\vec{e}_i \cdot \vec{e}_j = \widehat{\sigma}\,\vec{e}_i\,\vec{e}_j = \eta_{ij} = 0.$$

We showed that

$$\eta_{ij} = \delta_{ij}$$

when the basis is orthonormal.

Index